PIAGET IN THE CLASSROOM

PIAGET IN THE CLASSROOM

EDITED BY

MILTON SCHWEBEL
&
JANE RAPH

Basic Books, Inc., Publishers

NEW YORK

Frontispiece photo by Andrew I. Schwebel

Library of Congress Catalog Number: 72–89190
ISBN: 0–465–05751–9 (cloth)
ISBN: 0–465–09728–6 (paper)
Printed in the United States of America
DESIGNED BY VINCENT TORRE
10 9 8 7 6 5 4 3 2

THE AUTHORS

Beverly Birns is Associate Professor of Education, State University of New York at Stony Brook, Long Island, New York. Dr. Birns also maintains her appointment at Albert Einstein College of Medicine as Visiting Clinical Associate Professor, Department of Psychiatry, where she has been engaged in infancy research as well as research on the relationship between cognitive development in young children and social class. Prior to the above work, Dr. Birns had been a student at the Jean-Jacques Rousseau Institute in Geneva, Switzerland.

Mireille de Meuron is Psychology Consultant at the Clarke Institute of Psychiatry of Toronto, Ontario, Canada. She studied several years at the Rousseau Institute in Geneva. Since then she has applied Piaget's method for clinical diagnostic purposes in a cognitive study and treatment program in the Gouverneur Health Services Program, Beth Israel Medical Center, New York City. During her stay in the United States she was a Lecturer at the Graduate School of Education, Rutgers University, New Brunswick, New Jersey.

Eleanor Duckworth is Senior Research Associate at the Atlantic Institute of Education, Halifax, Nova Scotia, Canada. She was enrolled as a graduate student and research assistant with Jean Piaget at the Institut des Sciences de l'Education in Geneva, Switzerland. Subsequently, she has served as translator of Piaget's North American lectures. She has worked in curriculum development and teacher training with a number of educational projects, among them the Elementary Science Study and the African Primary Science Program. Her work with teachers led to the making of a film sponsored by the Ford Foundation, *Learning about Thinking and Vice Versa.*

Mark Golden is Associate Professor of Education, State University of New York at Stony Brook, Long Island, New York. He has been involved in studies of social class influences on cognitive and personality development in infancy and early childhood. At Albert Einstein College of Medicine he served as Assistant Professor of Psychology and principal investigator in a series of studies on the transition of infants from sensori-motor to verbal intelligence.

Howard E. Gruber is Professor of Psychology, Institute for Cognitive Studies, Rutgers University, Newark, New Jersey. He was a Research Fellow, Centre d'Epistémologie Génétique in Geneva, Switzerland, for one year and has made other visits there and had numerous interviews with Jean Piaget for a study of the development of Piaget's own creative thought. Dr. Gruber's work has involved the areas of perception, thinking, history of science, education and child development, psychology, and social action. Included among his numerous publications is a forthcoming book, *Darwin on Man: A Psychological Study of Creativity*, in collaboration with Paul H. Barrett.

Constance Kamii is a National Institute of Mental Health Research Fellow, International Center of Genetic Epistemology and the University of Geneva, Switzerland. From an early interest in diagnosis and remediation she has shifted to a focus on prevention of school problems. In this connection she has provided a leadership role in applying the implications of Piaget's theory to curriculum development in preschool and day-care classrooms. She has been affiliated with the Ypsilanti Early Education Program and a similar program in a day-care center at the University of Illinois at Chicago Circle.

JANE RAPH is Professor in the Department of Psychological Foundations, Graduate School of Education, Rutgers University, New Brunswick, New Jersey. She is presently coordinating a teacher preparation program in an inner-city day-care setting. She has had a long-time interest in the relationship of thought to language and speech in young children.

MILTON SCHWEBEL is Dean of the Graduate School of Education and Professor in the Department of Psychological Foundations at Rutgers University, New Brunswick, New Jersey. He has worked extensively in the area of educability and its development, and is the author of *Who Can Be Educated?* (1968). He has recently completed a study on logical thinking in college freshmen.

HERMINA SINCLAIR is Professor of Psycholinguistics, Ecole de Psychologie et des Sciences de l'éducation, Université de Genève, Genève, Switzerland. She has been a student, assistant, and colleague of Jean Piaget's, and her special interests have been in the areas of cognitive development and in developmental psycholinguistics. Her research in collaboration with others in Geneva covers a broad spectrum and has been reported extensively in both European and American scholarly journals and books.

GILBERT VOYAT is Associate Professor of Psychology, the Graduate Faculty, City College of the City University of New York where he is associated with the graduate program in clinical psychology. He was a student, research assistant, associate, and instructor for several years at the Centre d'Epistémologie Génétique in Geneva, Switzerland. His research includes the fields of identity, mental imagery, memory, and causality. In this country he has been affiliated with the Office of Health, Education, and Welfare's Indian Health Service in the Department of Mental Health. In this connection he has conducted various comparative studies and workshops among Sioux Indian teachers and children in South Dakota, and among the Navajo Indians in Arizona and New Mexico.

DAVID WICKENS is Associate Director of Follow Through at Bank Street College of Education, New York City, New York, with primary responsibility for the development of techniques for the analysis of educational programs relevant to teacher education. He has been a teacher at Bank Street, the University of Chicago Laboratory School, and public schools in New York City, as well as Director of Training at the preschool center at George Peabody College for Teachers. He has also designed various curriculum approaches for children and parents.

FOREWORD

The very useful work which I have the honor of prefacing possesses two qualities which are unfortunately less widespread than one might expect in studies of this kind.

First, the text has a concrete approach which goes beyond those educational theories drawn deductively from psychology. Instead of being restricted in scope to the statement of general concepts, the various chapters provide details of factual situations and teaching practices in the school environment. More specifically, the authors of this book convey their conviction that the field of experimental pedagogy must remain autonomous while utilizing the findings of psychology (in the same way medicine utilizes psychology), and that all hypotheses derived from psychology must be verified, through actual classroom practices and educational results, rather than merely based on simple deductions.

And second, the authors demonstrate a thorough comprehension of the role played by actions in the development of children's intelligence and knowledge; actions such as these refer to the manipulation of objects and the significance of such experimental and active manipulations in the understanding of the transformation of objects. In contrast with this approach, it appears that many educators, believing themselves to be applying my psychological principles, limit themselves to showing the objects without having the children manipulate them, or, still worse, simply present audio-visual representations of objects (pictures, films, and so on) in the erroneous belief that the mere fact of perceiving the objects and their transformations will be equivalent to direct action of the learner in the experience. The latter is a grave error since action is only instructive when it involves the concrete and spontaneous participation of the child himself with all the tentative gropings and apparent waste of time that such involvement implies. It is absolutely necessary that learners have at their disposal concrete material experiences (and not merely pictures), and that they form their own hypotheses and verify them (or not verify them) themselves through their own active manipulations. The observed activities of others, including those of the teacher, are not formative of new organizations in the child.

This essential principle, which has so often been misinterpreted, has been well understood by the authors of this book and they are to be commended for their insight. Undoubtedly, such a volume will help correct some fundamental errors and open the way for the implementation of a really active pedagogical practice, instead of retaining under the guise of a new language the mistakes of classical educational theory, which reduces the learner's role to looking and listening instead of acting himself, or which (and this is almost the same thing) replaces objects with audio-visual representations without concern for the fundamental role of spontaneous manipulations.

J. PIAGET

PREFACE

Nobody doubts that education in the schools needs improvement. But what can one do about it?

Some answers are obvious though difficult to realize. The schools of the world could instantly be better places for children, without resort to intervening studies, if children were well fed and housed, their parents were employed, and they and their families were treated with respect both in the schools and in the community.

Some answers are still very much beyond our reach. In our daily work as teachers or as researchers, we struggle for greater understanding of how children develop and learn and of what teachers ought to do to facilitate these processes. The answers must be practical ones, useful to teachers working in any of the schools of the world, in cities, suburbs, and villages. Practical solutions are found, as we know, not by random shots in the dark, but by systematic tests of the applicability of theories that make sense, that is, theories that appear to be effective in explaining cognitive development. As faculty in a graduate school of education, our daily preoccupation with the problems of the schools in the most urban state of the nation has led us, along with many others, to the theories of one of the fertile minds of the century. The lifelong work of Professor Jean Piaget provides rich opportunities to search for some of the necessary answers to our question.

We have applied Piaget's ideas to our own work in collaborating with a city school where, in the early stages, Edith Edelson provided the stimulation of her wisdom about young children. We also were engaged in studying logical thinking in college freshmen. The idea of this book emerged from the Rutgers Piagetian Conference of educators and investigators, first proposed by Robert Schwebel after he studied with Professors Piaget and Sinclair and worked as Sinclair's assistant. The Conference was organized to aid us in our work and to advance the level of conceptualization in teacher education.

We are indebted to the National Council on Early Childhood and to Frank Caplan for support of the conference. Participants included, in addition to the authors of this book, Milly Almy, Harry Beilin, Lois Beilin, Edward Chittenden, Betty Conley, Hans Furth, Herbert Ginsburg, Felice Gordis, Edith Neimark, Robert Parker, Joyce Weil, and Klaus Witz. Constance Kamii helped plan the conference and Rutgers graduate students Lew Gantwerk, Sue and Ray Martorano, and John Webster provided necessary supportive services. Howard Gruber advised us about the manuscript. Eliane Condon, Ranney Childress, and Mireille de Meuron assisted us in translations from the original French of Jean Piaget.

We acknowledge with gratitude the literary competence of our authors and their cooperation in meeting deadlines. The typing assistance of Barbara Furmick and Marion Keller was invaluable.

We are especially thankful to Professor Piaget for helping to make more comprehensible the complexities of human development.

New Brunswick, New Jersey M.S.
1973 J.R.

CONTENTS

PART III

The Developing Teacher

Introduction

. . . if the aim of intellectual training is to form the intelligence rather than to stock the memory, and to produce intellectual explorers rather than mere erudition, then traditional education is manifestly guilty of a grave deficiency.

(Piaget, 1970, p. 51)

CHAPTER 1

BEFORE AND BEYOND THE THREE R's

Milton Schwebel and Jane Raph

Jean Piaget's work is important to education for the most obvious of reasons: He has deepened our understanding of the experience of childhood.

Although he addressed himself to the problems of education as long ago as the 1930s, was active on a UNESCO commission in the early 1950s, and published a new work on the science of education in the late 1960s, his efforts have been devoted to the study of the nature of knowledge and the psychology of the child rather than education per se. It is left to others, especially to those of us in education and related fields, to mine the rich ore of his work.

Such ore is there in abundance, almost in response to the yearning of teachers and educators. Sheldon (1867), the author of a book for teachers, wrote:

> For many years there has been a growing conviction in the minds of the thinking men of this country that our methods of primary instruction are very defective because they are not properly adapted either to the mental, moral or physical conditions of childhood. But little reference has hitherto been had to any natural order or development of the faculties or to the many peculiar characteristics of children.

The investigations that Piaget began after World War I seem to have been custom-made to these specifications, written after our Civil War. They have revealed a "natural order or development" of the child and they have focused on the mental and moral development as well as the physical (sensorimotor) factors related thereto.

The ore is indeed there to be mined for a science of education from the work of Piaget, from his collaborators and colleagues, and from others around the world influenced by that work. Of course, such a science should draw upon all theories and empirical studies that illuminate the many areas of darkness. The authors of this book have chosen, at this time, at least, to concentrate on the theories that emanated from Geneva. Our aim is to carry the application to education of Piagetian theory beyond generalizations and certainly beyond the level of cliché. This we seek to do with full recognition of the fact that there is no simple one-to-one relationship between a principle about the order of development of certain cognitive operations between, let us say, the ages of 4 to 8 and the way a first-grade teacher conducts her class. But we are sure that principles of teaching deduced from the knowledge of the child's intellectual development can significantly and qualitatively alter the behavior of the teacher and the nature of the experiences she arranges for the children.

Each author in his or her own way has written what is meant to help the teacher function more effectively. The intended help is in the form either of direct suggestions about teaching per se, or of new insights about how intelligence develops and modifies and some of the vicissitudes of that experience.

No one, least of all we editors, claims that this book offers any panacea, or is anything more than the attempt on the part of the authors and editors to find ways of reorienting our behavior in schools so that more attention is paid to cultivating intelligence and to presenting information in such ways that it is meaningful and consequently useful.

Does the School Fit the Theory?

The ideas in this book, whatever their merit, are worthy of consideration only when those who shape educational policy in a school system have certain kinds of goals. Of course, this applies to virtually any kind of theory, Piagetian or otherwise. Freudian, Pavlovian, or Rogerian theory is selected to inform the work of practitioners in accordance with their respective objectives. At least, the thoughtful selection of a guiding theory has to be done on the basis of goals on the one hand and of the practitioner's confidence in the capacity of a particular theory to guide his behavior toward goal achievement on the other. In the psychotherapeutic situation when a person complains about frigidity (or impotence), the therapist responds according to at least two factors. First, he starts with assumptions about the relative merits of the various theories explaining the person's symptoms and decides (1) whether to contend with unconscious conflicts that must be brought to consciousness, (2) whether to recondition the individual to react differently to sexual stimuli, or (3) whether to enable the individual to strengthen his self-concept and his will. A second factor is the effect of the procedure on the individual himself or herself, that is, the advantages of one or another theoretical orientation in connection with the long-term effects either one will have on the way the individual looks upon himself and deals with subsequent problems in life. Will she be a stronger person not only in contending with frigidity but also in discovering the influence of her unconscious ideas? Will he be a more effective person not only in dealing with his impotence but in learning the power of conditioning and reconditioning? Will she once again not only deal with her problem but become more effective by having considerable conscious control over her concept of self and her will to change her own destiny?

Because in actuality most therapists are committed to one or another theory (or philosophy), a decision like this is usually made at some point of one's career and applies to most if not all professional decision-making in connection with problems of people. The example is intended to convey the importance of these two factors in making decisions about choice of theories that guide behavior, for the same factors apply to the choice of a theory to guide the behavior of the teacher.

The first question is what does the school system wish to accomplish? When governments, often quite reluctantly, first offered elementary education to the general population, the intention was to give such people the mastery over the three R's, to enable them to give their due to God and to their employer in the early Industrial Age. In some instances the leaders of this movement were quite explicit about not wishing the pupils to go beyond that level of education in the fear that they would then become dissatisfied with their lot.

School systems that have comparable objectives, even if at a somewhat higher level, ought to ignore this book. Piaget's work is almost in a world apart from the kind of school that looks for a mindless mastery of the techniques of reading, writing, and arithmetic. This statement neither belittles the value of those skills nor suggests that the application of views about cognitive development is in conflict with success in mastering skills. Not at all. To the extent that Piaget's theory or other theories of mental development do have significant application to the educational process, they are inappropriate in school systems that value orthodoxy in thought and behavior and highly disciplined technical skill in contrast to creative and imaginative productivity.

The school system must also encompass a leadership with confidence in the ability of its teachers to organize and to plan classes in accordance with the needs of the children in a given year. A theory such as Piaget's is inappropriate in a highly centralized school setting in which, for example, the minister of education in the capital city looking at his watch at 10:20 A.M. on a Tuesday morning knows that all 9-year-old children are learning the rudiments of long division with four digits in the dividend and two in the divisor. No educator or teacher who wants to follow a carefully prescribed syllabus that spells out each task to be done by the hour, each book to be used, each exercise to be practiced, and each test to be taken will find much use for this book. That is not to say that his educational objectives are "wrong"; they are simply different from those implied in this book. We are addressing ourselves to a different kind of school, a different kind of climate, a different kind of teacher, and different kinds of children—one might say—"turned-on teachers" and "turned-on kids." True enough, that is a slogan but it refers to the basis of this book: it should free teachers to be imaginative and innovative in helping children use the curiosity that they exhibit from their youngest days and the materials and experiences the teachers provide for them to continue their discovery and construction of the real world—that is, to continue the process of developing intelligence.

One either does or does not have confidence in the good sense of human beings. One either has or lacks confidence in the capacity of teachers, given the necessary training, to use their intelligence to plan appropriate experiences for groups of children and individuals within that group; and confidence in children to use those experiences to the advantage of their developing intelligence and knowledge. Without such confidence boards of education and school administrators maintain the kinds of control over the classroom that discourage invention and creative behavior.

Those who remember the great blackout in the northeastern part of the country in the mid-sixties recall the good feelings that were expressed about people. Somehow or other people were at their most spontaneous and ingenious. Social philosophers, east and west, struggling with the question of alienation in our society, seek to find how human beings can be turned on, on a continuing basis, by life itself. This aspiration is an appropriate one to apply to the school life of children. Turned-on intelligence is what we try to convey in this book, even in those highly technical sections that appear to be remote from spontaneity, an appearance that belies the facts. This is not a labored "turning-on," but an organized system to prevent the turning off, to sustain the curiosity and persistent exploration displayed by the infant and the child, and to provide some of the natural conditions and natural-social problems that intrigue the child from the early months and that intrigue adults as well. The terminology of these sentences is frequently associated with a highly permissive culture of our day and is sometimes associated with the absence of discipline. It may therefore seem ironic that the assumption in our work is that from the freedom of the teacher to use her intelligence to plan a variety of experiences for children—perhaps each different from the other from year to year and from class to class—emanates an organized and systematic mind capable of disciplined thought and, we hope, lofty and artistic creations.

One can create a rough test of confidence to enable a school system—and a teacher—to determine whether the Piagetian type of theory is an appropriate one. In addition to the major items discussed above, namely, the type of child one hopes will develop as the result of educational experience and the kind of teacher-role one intends to define, there are several other criteria. Can an active process of education be tolerated in which children are not necessarily sitting in an assigned seat during much of the day but are interacting with the materials in the classroom and working with the learning from one another? Can group activity be tolerated so that there is encouragement of projects jointly conceived and executed and for which the rewards and reinforcements are made to the group rather than to any individuals? Perhaps another way of asking this is whether cooperation can be tolerated, in fact encouraged and rewarded, in contrast to competitiveness, for through the process of social interaction children get a clearer understanding of the world by having some of their misconceptions recast through experiences.

Can incorrect language forms be tolerated and accepted, especially when the children, with the teacher's encouragement, are making discoveries or are learning new ideas? Conversely, is the teacher aware that correct language or the correct answer to a question may not always reflect a child's real understanding? Not only are 4 year olds taught by the teacher to reply in rote fashion that the month of March which comes in like a lion will go out like a lamb being given false information, but their ability to respond verbally is unsupported by their actual knowledge pertaining to the whole concept of months, the notion of beginning and end, or the ability to sequence with meaning the progression of a month made up of thirty-one days. Four year olds are still grappling with such time concepts as *before* and *after; today*, *tomorrow*, and *yesterday;* and *first* and *second.*

Can the teacher accept "wrong" answers that are wrong in the absolute sense but appropriate and normal for a child at a given age or must the teacher correct the child, thus leaving an imprint on the child who cannot possibly comprehend the correction, the imprint that the child is inadequate to the task? For example, the 4 year old who reasons that weight varies with size, that large objects are inherently heavier than small objects, is unlikely to benefit from the teacher's pointing out his faulty thinking and providing him with a correct explanation. Of greater value to the child will be many experiences that permit and encourage him to compare sizes and weights. Then he may begin to draw for himself conclusions that are in accord with his observations and that enhance rather than diminish his sense of self-worth, still very tenuous at this time.

In some ways none of these ideas is particularly novel. We merely ask whether a given school system and teachers can tolerate the conditions, nay can welcome and initiate the conditions, that good teachers at all times have used to foster the intellectual and social development of children. The difference is that we have a theoretical design and some empirical data supporting its validity to explain the stages and processes whereby the child's intelligence presumably develops.

All Mind and All Emotions?

One indisputable objective in the education and upbringing of children is the development of logical thought. Abstract thinking represents the great advance in the evolution of living species.

During an epoch of alienation when some specialists in mental health and education respond by de-emphasizing the cognitive as if its development were at a cost to the affective, it is necessary to point to the interdependency of the two. The nature of the period of concrete operations in the child from approximately age 7 to 11 or 12 restricts his interpersonal and social behavioral choices to reality, to that with which he has had concrete experience. A universe of possibilities opens up to the adolescent who has mastered understandings at the concrete level and now enters the stage of formal operations, whereby through hypothesis and deduction he can anticipate conditions different from those in immediate experience. New mental structures help liberate a person from his past and inaugurate "new activities which at the formal operatory level are mainly oriented toward the future" (Piaget and Inhelder, 1969).

In his work on the development of moral judgment, Kohlberg (1969) has found a high positive relationship between the cognitive level and the level of moral judgment of the individual. This is not to speak of a causal relationship but to suggest that abstractions such as justice, equality, and the golden rule are acquired about the same time as formal thought and are probably dependent upon some of the same mental structures and mechanisms.

Although it may seem obvious, the existence today of a proaffective and an anticognitive outlook demands clear specification of the interrelationship of the two and of the importance of formal or abstract thought to both of them. How one comprehends others and oneself has obvious implications for one's emotional experience of self and others.

Parallels: Evolution of Science and the Development of the Human

A striking parallel may be found between the onset of the formal stage in the life of children and certain stages or periods in the history of science. These two—the advanced thinking of a child and certain advances of science—have much in common. Let us first consider development of a child's thinking. Sinclair's research (discussed in Chapter 3) shows that the attempt to facilitate the movement from one stage to another is directed at bringing out contradictions in the thinking of the child. This effort appears to bear fruit only when the child is close enough to the point of transition to recognize the inconsistencies between his thinking as manifested in his proposed solution to a problem and reality when he applies his proposed solution. For example, he is at the point to recognize that while the tall, thin beaker appears to contain more water than the short, wide one, the water poured from the tall one to another short, wide one is the same amount as in the other short one. The tall one still seems to have contained more water, but the child is now perplexed by the inconsistencies and is on his way to controlling variables and acquiring knowledge of the conservation of liquids. As another example, when the child is in process of transition, he recognizes that the 10 marbles compactly arranged are in fact as numerous as the 10 occupying a larger space. He comes to this conclusion by recognizing the contradiction between his notions about space and number, which he can resolve in this instance by controlling the space variable and focusing only on quantity. The point to be made here is that development occurs when the child finds the inconsistencies and is ready to cope with them by inventing new intellectual structures.

The importance of awareness of contradictions is emphasized also by Kuhn (1962). His description of the relationship between "normal science" and the scientific revolutions that lead to a qualitative change in scientific thought seems to reflect Piagetian language. He defines normal science as ". . . the activity in which most scientists inevitably spend more of their time . . ." (p. 5). It is based on the assumption that scientists as a group know what the world is like. The counterpart in human development is the individual's necessary assumption that he knows what the world is like; at least his conceptions of the world are all he has and thus all he can rely upon to interact with it. Because he is limited to what he has he cannot at this point in his development recognize certain contradictions. For example, the fact that the water in the tall, thin glass is poured into the short, wide one and then back again, does not alter his conviction that the quantity is greater in the tall one. He does not yet see the contradiction. Kuhn says that normal science often suppresses fundamental novelties, but not for very long. Change occurs either because a normal problem resists solution or because during research *some anomaly is revealed*. When such an anomaly occurs, the sciences begin "the extraordinary investigations" that lead to "a new basis for the practice of science." These great episodes he calls the "scientific revolutions," for they are "the tradition-shattering complements to the tradition-bound activity of normal science" (Kuhn, p. 6).

The infant and the child also have their periods of "extraordinary investigations." For a time they too are on a plateau, conducting their own "normal science," solving problems with the limited repertoire at their disposal, and suppressing, or, better yet, being unaware of some problems or of the anomalies in some of those they think they solve. Then comes a time when these anomalies are apparent, or so apparent they cannot be ignored. The child's investigations are his interactions with the physical and social environment aimed at resolving the contradictions. These are his revolutions. And revolutions they are in the life of an individual when he discovers that the amount of clay is unchanged even as the form of the clay is changed, or that the water poured from one vessel to another likewise is the same in quantity though it reaches a higher point in one glass than in another. The discovery of conservation is indeed a scientific revolution in the life of the child.

Each revolution in science has led to a shift in the kinds of problems that scientists can attack and also a change in the standards of an acceptable problem-solution. For example, after Copernicus, after Newton, after Lavoisier, and after Einstein changes occurred in both of these.

No less so for the child. Compare the kinds of problems that attract the attention of those aged 5 and 15: e.g., at 5, how to put dolly to sleep; at 15, how to plan the junior prom. Also, the quality of the solution that is appropriate to the age group is very different: e.g., the definition of roles in ball games at 5 and 15.

The practical problem for the teacher centers on the periods of both normal science and scientific revolutions in the development of the children. First comes the question of the experiences or tasks that help a child consolidate his gains during the "normal" period, and next, those that help facilitate the revolutions, the giant steps to the next stage. For both of these an active process is essential.

An Active Process

The reader would be justified in looking askance at this topic heading, for so much has been said about the subject over so long a period of time probably with few beneficial results. Nevertheless, it is worthwhile to take a fresh look at it, to consider further why and what kind of activity is desirable.

Activity per se is not a factor in intellectual development. Even the most vigorous of calisthenics cannot be credited with promoting mental growth, except in the remote way in which the positive functioning of the total organism benefits every aspect of functioning. The facts that children move about the classroom, that seats are not fixed, and that children even hop, skip, and jump do not make the active process educative.

The key is quite obviously the nature of the activity. As a prototype a 12-month-old child interacts with an object, a 2 × 2 red and yellow plastic block. He squeezes it between his two hands. He moves it up to his mouth, bites at it, and pushes it away so that it falls to the floor. He picks it up and repeats the previous steps. He picks it up, clutches it, and hammers it on the floor. Then he throws it. This time it drops not because he pushes it away from his mouth but because he has thrown it. Over a period of weeks he repeats these actions in many different sequences involving an increasing number, variety, and complexity of perceptual-motor coordinations. During this time he is (1) discovering some of the properties of the object through his eyes, hands, mouth, ears, and muscular action—discoveries limited only by the mental level of the 18 month old and, in particular, this 18 month old; he is at the same time (2) inventing coordinated actions that are independent of the particular physical object (the block) though dependent on some such object. The coordinated actions involve perceptual and motor behavior. The child looks in the direction of the block, sees the block, extends his arms in the direction of the block, adjusts his movement to the proper distance, grasps it, clutches it, lifts it to his mouth, sucks it, etc.

The test of the effects of these experiences lies in the child's subsequent behavior. Has he acquired a representation of the physical object so that he would seek it out when it was removed and would recognize it upon its return? Does he use the same kinds of coordinated actions that he presumably constructed in his encounters with the object when other situations stimulate such coordinated responses, like the placement of a recognizable sweet at a similar distance from him?

The important role of activity becomes less obvious and more subject to superfluous interpretation with each older group of children. As early as about 1½ to 2 years the child's representation reaches a higher level, at which time he can use some signifier to represent an object or an event. The effects of his action are apparent only to the conscious observer. Furthermore, his actions at the point of acquiring the representation, as in the case below, may have had no motor action connected with it at all.

Piaget and Inhelder (1969) give some examples of the signifier in representation, one of which we will describe. An 18-month-old girl engaged in the following deferred imitation: She observes a playmate become angry, scream, and stamp her foot. She has never before observed such behavior. An hour or two later when her friend has departed the girl imitates the playmate's behavior, though in a state of laughter rather than genuine anger. Deferred imitation like this constitutes representation, that is, the brain has assimilated the event to the extent that the event can be signified, can be represented though not reproduced.

The process of developing a matrix of fundamental knowledge during childhood is a long one, requiring active engagement over periods of years. It may not be true that time *qua* time is central, but time can hardly be partialed out of the developmental history of the child. Take the child's experience with an apple as an example. During the sensori-motor period it is an object not unlike the block except that its shape requires different grasping behavior, and its scent and edibility evoke different oral activity. Consider the innumerable experiences with this fruit before the child completes all of the assimilations and accommodations necessary to recognize it by sight, by feel, by smell, and by taste; to differentiate it from other round, colored, scented, edible fruits; to have representations of it that are imitative or pictorial, that is, figurative, and eventually symbolic; to place it in a class of fruits, and to relate this to other classes of fruits, and those to categories of food; to understand the relationship of the apple to a tree, the tree to an orchard, and that to the natural (e.g., sun, rain) and social (e.g., farmer, fruiterer) forces that bear upon them.

The child needs time to make these discoveries. The adult can be helpful in providing and arranging experiences including the social experiences of eating and discussing apples, but he cannot inculcate what the child cannot accommodate, i.e., he cannot substitute memorized facts for the discoveries that the individual makes in working with the concrete object or the various symbols thereof. Memorization of meaningless material is on a par with learning nonsense syllables as far as development and learning are concerned. There may be conditions in which memorization and drill are desirable in the education of a child; however this process is not the same as the child's internal construction of reality, as his assimilation of knowledge of the world, or as his concept formulation—knowledge that helps in the comprehension and control of the environment. Such memorization may have some immediate ego value for a given child or for a teacher, but it is not usable learning.

The matrix of conceptions based on active engagement of the child in the first years of life also comprises the precursors to reading. Reading is an activity that requires the mental capacity to represent objects with conventional signs that are distinct from the objects for which they stand. The means through which the child develops this mental capacity evolve gradually from such an active interaction with the concrete world we have described. The more meaningful and varied the encounters the child has had with the real world the better will be the "thinking foundation" on which learning to read is necessarily based (Furth, 1970). These encounters have a cer tain order to them as the child's growing and differentiating abilities meet expanding and more complex features of his environment. Although the idea of reading's being related to experience has been around for a long time, there is still a tendency to assume that the 6 year old entering first grade should have sufficient experiential background to grasp the meaning of words in charts, workbooks, and readers. Piaget's careful observation of how knowledge develops in children has given us a better understanding of the nature of the sequential experiences a child has had prior to age 6 which enable him to approach the goal of school learning. Piaget's system also provides us with clues to use in discovering a child's level of thinking and, accordingly, the adequacy of his preparation for reading.

Six or seven kinds of encounters leading to reading can be charted. First, as mentioned previously, during the first two years of life, imitation is a main source of learning. The young child uses his body to represent objects or expresses his mental image of an object by his actions. If shown a spoon, he pretends to eat. A pan that turns up in his toy chest becomes a basis for his pretending to cook. These actions and meanings are private to him, based on his observations, but suggest also that he has a mental image of what he is doing, that he is connecting an every-day experience he has observed with his own actions—he is internalizing a concept. Next he becomes able to substitute one object for another. He uses blocks for cars, chairs for a train, tinkertoys for food, etc. Utterances become representations, too. "Ding-a-ling" means a telephone, "quack-quack" a duck, "toot-toot" a train, etc. As he begins to engage in sociodramatic play with other children, not only does he imitate his mother's cooking at the stove, tending the baby, or going to the store, but he *becomes* mother (in his thinking) in relation to the other children with whom he is playing. This may mean sitting down with the visitors for coffee, sending father off to work, or preparing the children for school. Sociodramatic play is considerably more advanced in thinking and in social action than was the simple imitation of an action (Smilansky, 1968).

In a next encounter, which may occur in sequential, step-wise fashion following the ones just described or may overlap previous mental constructions and anticipate new ones, the child forms objects out of clay. When earlier he may have built a single tower of blocks or arranged a random collection, he now purposefully joins block buildings together to make a city, a bridge, a farm, etc. From there he proceeds to draw representations of what he has seen on a trip, what he has built, or what he has observed at home. Identification of pictures occurs next. Although some children are able to identify pictures at a very early age, it is only after subsystems have become integrated representations in their internal construction of fundamental knowledge that youngsters are able to interpret flat, one-dimensional pictures and coordinate their interpretation of the picture with their previous experience. For example, a child may be able to identify a picture of a duck by its appropriate label, but he has to have had some experiences with ducks, preferably repeated experiences, before he can derive the full meaning of "duckness" from the picture. If he has held a duckling in his hand, fed ducks at the pond, listened to their sounds, watched them swim and dive, and observed their waddling walk on the shore of the pond, then the representation of the duck, and eventually the word "duck," is understandable. In other words, pictures are an important bridge from subject or action to the sign or word, but are not sufficient in themselves.

Note should be made here that the skillful teacher may well begin by backtracking, using Piaget's sequence as a guide before confronting children with "foreign" symbols and signs, even pictures. Sigel (1971) demonstrated, for instance, that lower-class children, when trained in classification skills applied to objects, did show increased competence in grouping three-dimensional items, but did not improve in ability to classify or group pictures. He pointed also to the need for studying the salience and relevance of classification ability to cognitive competence and to the learning of reading and numbers.

Thus we see that the first years of a child's life are made up of a long, busy, active history in which he, beginning with his own observations and imitating them, acquires a differentiation of understanding through activity. He learns from his interaction with children and adults. He learns from his repeated experiences—pouring, transferring various quantities of liquids and solids from different size and shape containers to others, stretching objects, bending, stacking, building, fastening, floating, tearing, cutting, pasting, coloring, painting, sociodramatic playing—the vast repertoire of actions on objects that enchant him and hold his attention. The knowledge in these activities comes from introducing a *relation* or a *transformation* between two objects and finding out the effects of such transformations or relations. This history of the child's development of structures becomes for him the forerunner of his meaningful comprehension in the reading process.

In one of the author's current investigations of logical thinking in college freshmen, some able students gave evidence of meaningless memorization they had engaged in with the obvious encouragement of well-meaning teachers. For example, one scholarship student, a physics major, was puzzled by his difficulty with the balance beam. This problem, described by Inhelder and Piaget (1958), requires the subject to explain the principles that account for the balance of unequal weights at unequal distances from the fulcrum. The subject is free; he is in fact encouraged to experiment with the materials and needs no prior study of physics or any other subject to discover the relationships between different sized weights and notched distances from the center. The student in question recalled almost immediately a formula he had learned in high school and was happy at the speed with which he thought he had dealt with the problem. His satisfaction turned to frustration when he was asked to apply his formula, that is, to predict the unequal weights and the locations that could be used to achieve balance. He could not get his formula to work. He had memorized it well but not understood the principle of balance; consequently he was measuring distance from the edge of the beam rather than from the center.

When this student was given the same problem eight months later, he responded in the same way. He could not resolve the contradiction between what memory told him was the correct answer and what experience told him was not. The second time he was a subject in a pilot intervention phase of the study. At the outset he had been asked to draw four straight lines to connect the nine dots in the figure below. He

• • •

• • •

• • •

struggled with the problem, and, as one who had always been highly successful in tests, he experienced a high degree of frustration. When the experimenter asked for the portion of the figure and field he was working with, the student delineated a square bounded by the eight outside dots. The experimenter suggested that he was stuck with his mental set—"stuck in a rut"—and with that the student proceeded to solve the problem.

Later, when the student was stymied by the balance problem, the experimenter suggested that he was again imprisoned in his own set. The student started afresh and discovered for himself the phenomenon of balance that the formal study of high school and college physics had not taught him. Here, in the setting of an experiment, at the suggestion that he break out of a mental rut, he freed himself to experiment with the object itself and to find the principles of its behavior.

A teacher in a middle school had the children work at building a model of the sewage system in the community. The children under his guidance participated with interest in the construction process, with small groups responsible for segments of the system. When the elements were completed and the teacher had put them in order, he moved the children from the active process of constructing separate segments to the passive process of memorizing the functions of each and their relationship to the whole. The teacher would ask a question, a few hands would go up, and a child would give a response that most of the children obviously did not understand and sometimes even the child responding could not have explained. Passages had been memorized from an outline book prepared to go with the sewage unit.

The teacher seemed to enjoy this recitation experience no more than the children but was not free to function any other way, probably because he knew no other way. Many in this position try their best to arouse and hold the interest of the children by following the very pattern he used: an active process to introduce the unit, almost a form of "fun and games" to get them hooked, and then the serious business of "learning."

The freedom to act upon the world and to construct reality

is both the aim and the process of education. It is not "learning by doing," now a cliché that justifies virtually any kind of activity as if the learning were in the act of doing. It is different from the ability to verbalize a correct answer or to obtain a correct solution. Learning requires comprehension: The individual must be able to assimilate new knowledge if it is to be usable. That is, knowledge must be assimilated—incorporated into a system that can accommodate it—for it to be meaningful.

Teachers cannot talk students into comprehending conservation or proportion, democracy or equality. They can lecture and force memorization by the use of reinforcements, but by virtue of that they are unlikely to develop the operations that are associated with the assimilation of such concepts.

It would be erroneous to say that memorization or even drill has no place in the teacher's repertoire. In foreign language study the student does help himself by learning the meaning of *l'homme* or *der Mann*. It is helpful for him to know why these are equivalent in meaning to their English counterpart, but that is not essential to his purpose nor does time make it possible in every instance. He learns by association that man is *l'homme* and vice versa. He relates London with England and Paris with France. Yet he learns the meaning of a capital city only through a long process that cannot be replaced by the act of lecture or by repetitive recitation. It takes time, it takes involvement, it takes action.

Much of our instructional planning is based on the notion that for the teacher to state some fact or principle is to teach it, and for the student to be able to restate it means it is learned. This idea permeates our experience at every level, especially in high school and college. Thus teachers cannot be faulted for perpetuating what they have experienced, nor students for performing what is expected. Piaget, however, has observed that language itself is not a necessary element of operational thinking, particularly in the young child who indeed speaks a language of his own. It may be helpful to the reader to consider some ramifications of the role of language in intellectual development.

The Role of Language

We live in a highly verbal society. A network of tests of achievement, intelligence, and even creativity that share an emphasis on vocabulary and standard English grammar is used as the basis for judgment of a child's academic level and potential. As a consequence the pupil's ability to supply the correct definition, give the right answer in response to a teacher's question, or circle the correct statement on an objective quiz is all too often equated with understanding. These are not necessarily the same. Piaget holds that the sources of thought are to be found not in language, but in the preverbal, sensori-motor actions performed and experienced by the very young child during the first two years of life. The infant, for example, directs his activity first to manipulations and then toward his own satisfactions. Later he will search for explanations and will want to tell others. Language shows his knowledge of objects, events, and relationships rather than his reaction to these (Sinclair-de-Zwart, 1969). This proposition is the reverse of the long-standing one in language learnings, something to the effect that the "word is the thing," that is, if we give a child suitable verbal training, he will have the tools with which to think, reason, and to interpret his experiences (Vygotsky, 1962; Bereiter and Engelmann, 1966).

In strong contrast to the emphasis on verbal proficiency, Piaget points out that logical thinking is primarily nonlinguistic, derived from first imitating actions and then performing actions. Piaget does not deny that social interaction and the resulting language acquisition are important. Verbalization may sharpen contradictions in the child's thinking and help bring about one of the "little revolutions" that propel him into a higher level of thinking. Language, according to Piaget, permits the child to evoke absent situations and liberate himself from the restrictions of the immediate, so as to extend and deepen his understandings. But Piaget is consistent in claiming that language alone does not account for the development of intellect.

Evidence for Piaget's position is found in the work of Furth (1966), who showed that deaf children have ability equal to normals when given a test of use of logical symbols, whereas blind children (deprived of opportunities for imitation and learning about relationships through actions on objects) solved the same type of tasks on an average of 4 years later than did normals (Hatwell, 1960). Sinclair-de-Zwart (1969) additionally demonstrated that children who were unable to conserve on a series of tasks, when taught the appropriate terms used by children who had succeeded on conservation tasks, still did not grasp the principle.

The meaningfulness of this concept that language reflects thought is profound, if not overwhelming, in its ramifications for curricular change. It posits first of all that the foundation for development of mental activity in a young child is the recognition of the importance of his potential as an active doer rather than a passive recipient of the wisdom of others. It de-emphasizes the role of the teacher as an explainer, director, or imparter of information. It casts some doubt on how productive it is, in the long run, to rely on verbal answers from students as an indicator of their grasp of relationships when, in fact, their verbal responses may tell us little about their conceptual understanding. It suggests further that our practice, in the recent preschools for inner-city children from poverty backgrounds, of emphasizing names of objects, recognition of different attributes of objects, and the correct verbal answers to an abstract type of teacher question may not be as productive in helping the child develop a grasp of a new phenomenon as his actual experiences with the phenomenon. It stresses, in addition, the importance of helping a child develop representation (mental symbols, visual and auditory images, etc.) from action and giving him ample opportunity for performing operations as a basis for the major cognitive tasks of the school-age child, that of mastering classes, relations, and numbers, and becoming able to conserve quantities. Since the acquisition of ability to think operatively occupies a key role, according to Piaget, in the thinking of the school-age child, it will be described next, occurring as it does in the normal course of development following the child's conquest of the object through sensori-motor actions and the conquest of the symbol during the preschool years.

The Nature of Concrete Operations

Concrete operations in the development of a child's thinking occur as he moves out of the preoperational stage associated generally with ages 2 to 7 years and prior to his ability to handle formal propositions beginning at about the ages of 11 to 12. As indicated previously, this attainment occurring around 7 or 8 years of age is considered a landmark achievement. Its importance lies (1) in its current contribution to the organization of mental actions in operational thinking capacities of the individual as applied to concrete objects; (2) in the antecedent experiences that have prepared the child, as it were, to construct understandings of reversibility, reciprocity, and constancy; and (3) in the child's grasp of relationships between parts and wholes, the structuring of which enables him later on to perform the mental functioning known as formal operations. The teacher's question may be, "How is operational thinking crucial in the child's development of intellectual structures?" or "What are some characteristics of the process as demonstrated in the school-age child's thinking?"

GROUPINGS

Piaget distinguishes grouping as the principle from which stem classification, seriation, conservation, number, and space understandings. He points to groupings as being essentially of three kinds: (1) those that pertain to identity or equivalence; (2) those that refer to the logical system of classes wherein two classes may be included in the other, or may partially overlap, or may be mutually exclusive (Baldwin, 1967), e.g., all robins and sparrows are birds; birds include robins, sparrows, ducks, and other species; the class of birds contains birds that swim and those that do not swim; and (3) those groupings that refer to relationships between the parts and whole of a concrete object or a collection of objects or persons, e.g., if Bobby is a brother of Ann, then Ann is a sister of Bobby. As Baldwin (1967) illustrates, in a series of four circles graded in size such that A is small, B is systematically slightly larger, C even larger, and D the largest, the child can see that if $A < B$, $B < C$, and $C < D$, then $A < D$. The child can also look at the four circles and reverse the process, namely, if $D > C$, $C > B$, and $B > A$, then, of course, $D > A$.

CLASSIFICATION

The ability to classify grows out of the child's experience with noting and acting upon resemblances and differences (Furth, 1966). The child first becomes able to put together elements on the basis of their similarities and differences. A 4 or 5 year old, given an array of blocks of varied colors, shapes, and sizes and asked "to put together what goes together," may start by placing a red square and a red triangle together. The observer may conclude (prematurely) that the child not only spontaneously recognizes the color red as different from others, but is also able to extract all reds from the array and organize them accordingly. Often, in fact, this child may proceed in quite a different manner in which, having placed the red triangle next to the red square, he then searches for other triangles, regardless of color, and then happening on a yellow triangle, he may turn to searching for yellow objects. Thus, he is using proximity, primarily, to guide his selection of what is related to what, producing what Inhelder and Piaget (1969) describe as graphic collections.

In concrete operational thinking the child's behavior progresses from simple, graphic collections to the ability to coordinate on the basis both of similarity and of relationships. The true basis of classification depends not upon recognition of perceptual likenesses, but upon the child's control of logical quantifiers such as *one*, *some*, and *all*. He also gradually becomes able to classify according to two or three criteria at once, e.g., these are all bottles, they are all made out of glass, they all have narrow tops; to shift criteria—becoming able to propose several ways to classify bottles, e.g., these bottles are translucent and those are not, these are small and those are large, etc.; and to hold to that single or multiple criteria in actually performing the classification act with the concrete objects.

SERIATION

Seriation is also a fundamental development in mental operations, a process that refers to the child's ability to arrange elements according to increasing or decreasing size, graded hues, graded heights, etc. A child first orders a nest of boxes from the smallest to the largest by chance arrangement or by trial and error. Slowly he begins to be able to coordinate the difficult thought (to him) that each box is smaller than the one into which it fits and larger than the one that is fitted into it. When a child orders a nesting correctly and understands how the nesting is accounted for, he can quickly and correctly replace one box that is removed from the series. If he does not yet mentally coordinate the relationships, he may have to begin again to assemble the series. He acquires coordination first by comparing two or three sizes, then by handling five or six sequential sizes; with concrete operational thinking he is not confused by a series of as many as ten or twelve sizes. In addition, he can look ahead and plan how to execute the seriation, say from largest to smallest, and with ease can reverse the process and arrange from smallest to largest.

CONSERVATION

Occurring almost simultaneously with the ability to grapple with groupings, classification, and seriation, the child also acquires construction of a system of regulations that enables him to compensate internally for an external change. When liquid is poured from a wide, short container to a narrow, tall container, he recognizes that the liquid may look like more in the second container but is the same, since (1) nothing has been added "to" or taken away "from" the amount of liquid; (2) if the liquid is returned to the original container, it will be the same amount; and (3) the narrow container is higher, but it is also thinner. These three processes of mental activity Piaget refers to as (1) simple additive identities; (2) reversibility by inversion; and (3) compensation or reversibility by reciprocal relationships (Piaget and Inhelder, 1969). The whole notion that certain aspects of an object remain invariant in the wake of substantive changes in other attributes comprises the essence of a universal in the growth of thinking (Flavell, 1963).

NUMBERS

The child's understanding of numbers is synthesis of the operations of class inclusion and seriation. The child during concrete operations acquires the ability to ignore differences in ascribing numbers, as *1* desk is equivalent to *1* chair and *1* pencil insofar as the number *1* is concerned. But he must also understand that numbers are seriable and that he should not count the same object twice in a series. He has to acquire, too, an understanding of logical mathematical relationships that a given number of objects when compared with an identical number of objects contains the same amount in spite of spatial arrangements that may make them look different.

SPACE

Our young operational thinker now knows where his house is in relation to school (and where school is in relation to his house). He becomes able mentally to fill the space between the two (the stores, church, other houses he passes enroute from one place to the other). His thinking is becoming at the same time more flexible and more stable. He is not taken in as easily by perceptual clues or verbal arguments. He literally "knows where he stands." He can coordinate two characteristics at the same time so that two sticks he recognizes as being the same length in a matched position retain their identicalness of length when one stick is pushed ahead of the other. Here he can compensate for the lack of identity on the left and on the right at the same time.

Flavell (1963) defines the general cognitive differences between the preschool and school-age child of which concrete operational thinking is the fundamental point as follows:

> It is simply that the older child seems to have at his command a coherent and integrated cognitive system with which he organizes and manipulates the world around him. Much more than his younger counterpart, he gives the decided impression of possessing a solid, cognitive bedrock, something flexible and plastic and yet consistent and enduring, with which he can structure the present in terms of the past without undue strain and dislocation, that is, without the ever-present tendency to tumble into the perplexity and contradiction which mark the preschooler (p. 165).

Of the many principles to be derived from Piaget's work which are applicable to a child's acquisition of knowledge in and out of the classroom, we have suggested here the pervasive contribution of activity that enables him to discover, interiorize, and build his understandings for himself, not merely copy them. We have further indicated that language is important only as it reflects, emanates from, and ties back into the total intellectual functioning of the child. Finally, we have described the nature of concrete operational thinking and its crucial role in the developing intellect.

Most important we have said that before and beyond the three R's—more precisely beneath the three R's—exists a human being who is not partly cognitive and partly emotional but in being and substance is both of these all of the time and indivisibly. As any teacher knows, whatever his effort to keep one of them—the mental—in the foreground during a class period, the other inevitably intrudes. In fact, many effective teachers will see to it that emotion in some human interest form, such as an anecdote or role-playing, does appear. The separation of the two is sometimes necessarily sought after in research.

There is a rhythm of human development beyond the three R's that probably encompasses them too, a rhythm that seems to parallel that of the sciences. These are the plateaus or periods of seemingly no progress, and the peaks that represent the great upward thrusts of revolutions. The recognition of conflict between what one thinks (e.g., the earth is flat) and what one finds (the ship did not sail off the edge of this flat earth) leads to new formulations, new knowledge, and advanced intelligence. The active and professionally developing teacher learns when and how to sharpen these contradictions, helping the child to move on to a qualitatively higher level of intelligence.

REFERENCES

Baldwin, A. *Theories of child development.* New York: John Wiley, 1967.

Bereiter, C., and Engelmann, S. *Teaching disadvantaged children in the preschool.* Englewood Cliffs, N.J.: Prentice-Hall, 1966.

Flavell, J. H. *The developmental psychology of Jean Piaget.* New York: Van Nostrand Reinhold, 1963.

Furth, H. *Thinking without language: Psychological implications of deafness.* New York: Free Press, 1966.

Furth, H. *Piaget for teachers.* New York: Prentice-Hall, 1970.

Hatwell, Y. *Privation sensorielle et intelligence.* Paris: Presses Universitaire de France, 1960.

Inhelder, B., and Piaget, J. *The growth of logical thinking from childhood to adolescence.* New York: Basic Books, 1958; London: Routledge and Kegan Paul, 1958.

Inhelder, B., and Piaget, J. *The early growth of logic in the child: Classification and seriation.* New York: Norton, 1969; London: Routledge and Kegan Paul, 1964.

Kohlberg, L. *Stages in the development of moral thought and action.* New York: Holt, Rinehart and Winston, 1969.

Kuhn, T. S. *The structure of scientific revolutions.* Chicago: University of Chicago Press, 1962.

Piaget, J. *Science of education and the psychology of the child.* New York: Orion Press, 1970; Paris: Denoel, 1969.

Piaget, J., and Inhelder, B. *The psychology of the child.* New York: Basic Books, 1969; (*La psychologie de l'enfant.* Paris: Presses Universitaire de France, 1963).

Sheldon. *Manual of elementary instruction.* 1867.

Sigel, I. The development of classificatory skills in young children: A training program. *Young Children 26*, 3 (1971), 170–184.

Sinclair-de-Zwart, H. Developmental psycholinguistics. In D. Elkind and J. Flavell (eds.), *Studies in cognitive development: Essays in honor of Jean Piaget.* New York: Oxford University Press, 1969.

Smilansky, S. *The effects of sociodramatic play on disadvantaged preschool children.* New York: John Wiley, 1968.

Vygotsky, L. S. *Thought and language.* Cambridge, Mass.: The Massachusetts Institute of Technology Press, 1962.

PART I

The Developing Mind

The essential functions of intelligence consist in understanding and inventing, in other words in building up structures by structuring reality.

(Piaget, 1970, p. 27)

EDITORS' INTRODUCTION

Show a child of four a short straight line and a longer zigzag one that starts where the shorter one starts but ends before it. The 4 year old says the straight one is longer. "Right?" or "Wrong?" The child says what a child his age comprehends. He is normal. If he is told his answer is wrong, he might learn to doubt himself, but he would not learn to deal with two variables simultaneously because he is not yet prepared to structure reality as the adult does. So the teacher must learn to recognize "wrong" answers as appropriate ones, in fact, as "right" ones for children of a certain age.

The study of the developing mind means the study of the natural ways in which a biosocial organism grows, learns, and matures. It means an ever greater understanding of the ways in which we can facilitate the process of development by improving our present forms of intervention or inventing new ones. This is not to say that we ought to be searching for ways of speeding up the process so that humans can skip childhood and go from infancy to adulthood. Piaget has twitted some American psychologists for wanting always to accelerate development, which is not at all the same as improving the social influences and experiences that are the social, psychological, and even physiological components of growing up, that is, providing the best possible circumstances for development.

If we are to improve our schools, we need to bring them more in harmony with the processes of development. This goal applies as much to the organization and climate of the school as a whole as to the social and intellectual character of classroom life. Practices that are antagonistic to the processes ought to be replaced by appropriate ones.

Part I tells of the developing mind. Sinclair begins her first of two chapters by pointing out an incompatibility that has to be considered. A major problem in the application of Piaget to education is that his work is concerned with the development of fundamental knowledge or intelligence in the child, whereas schools are interested in the child's acquisition of skills and information. These two are related because information is useful only as there is the underlying knowledge to make it meaningful; nonetheless, these two are fundamentally different. Schools concentrate on skills and information, not infrequently in the absence of knowledge, and measure the achievement of the child in those terms.

Fundamental knowledge is acquired not through the senses directly but through action upon and interaction with the environment. The drastic change in that knowledge—i.e., in childrens' thinking between the ages of 4 and 8—is in the acquisition of reversibility; children of 8 are able mentally to cancel out one action and perform another. For instance, a *long* toy *truck* can be stored in a drawer with *big* teddy bears and the like; it can also be stored with toy *cars* of all sizes. Or a boy of 4 or 5 who knows he has two brothers but says his brother *A* has only one brother, can by the age of 8 engage in the necessary mental transformations to see himself as one of the brothers.

Sinclair considers what it is that paves the way for reversibility—what earlier acquisitions prepare for its development. She traces the development of object-permanency and speaks of its nature to mean that the subject's actions and the object itself are "inextricably linked." She does the same for the qualitative identity of the object, i.e., that several different actions can be performed on the object without altering its identity (biting it, throwing it, etc.). Then comes the important development of quantitative constants whereby the child realizes that variations in shape, position, and the like do not change the quantitative aspects of the objects (number of marbles in changed arrays; amount of liquid in vessels of different shape, etc.).

Having dealt with the development of the constancy of objects, she next considers change in the structure of operations, i.e., the action structures that, she makes emphatic, are in reality dissociable from the objects. The period between 4 and 8 sees the child move from a functional dependency of one object upon another (a push causes the block to move) to the covariation of objects (when it is rolled, clay gets longer and also thinner); by 8 the child has reached covariation without the mediation of action, that is, understanding that getting longer is compensated by getting thinner.

Sinclair's final section is on the development of the symbolization processes and the influence on them of cognitive development. She shows how the child deforms information presented in visual and verbal manner (the chief modes of the classroom) in accordance with the knowledge of the child. What the child cannot assimilate in its material state, he deforms into an accommodatable form. Many a teacher who is not aware of that natural process goes blithely on his way believing that what is presented in class is—or ought to be—learned as such, and that a child's failure to learn it, as it is taught, is attributable to either cognitive or motivational deficiencies.

In Chapter 3 Sinclair relates her efforts and those of several colleagues to investigate the possibility of directed influence over the movement from one stage of mental development to a higher one. She first explains that development occurs as the child discovers links between events. The link cannot be imposed upon him; it occurs as a result of his interactions with objects and people. New behavior results from recombination of existing operations or action schemes.

Sinclair reports in fascinating detail a study that tried to arrange children's encounters with the environment in such a way that the children could be observed at the very time that they were recombining existing operations into new ones. Problems (and behavior) in three areas were studied: (1) geometry, the conservation of length; (2) logic, the problem of class inclusion; and (3) physics, a problem of the conservation of volume. In the study situation or in natural life it is the "active re-combination of schemes by the child himself which results in the acquisition of a new concept."

Sinclair's analysis of the transition from early to mid-childhood thinking reveals that time, and the activities of life that of course occupy time, are essential to change. There are no snap-of-the-finger alterations in mental functioning, as both the practicing psychotherapist and the practicing teacher can attest. That does not mean that time by itself, if that could be imagined, is sufficient—just that time is a necessary although not a sufficient factor. It is what we do in the interim, Sinclair suggests, that can influence perhaps both the quality and the speed of change.

Gruber, analyzing the behavior of an adult, Charles Darwin, demonstrates the significance of time and explains some of the mental operations that occupy periods that look like plateaus in the developmental topology. He reflects upon the reasons why significant changes in the structure of ideas take time. These include a modification of our vital systems, that is, of the very ways we experience the world. Changes of that character which are part of life for all of us affect our social relations in that we move from one to another way of structuring reality while others continue to structure it in the old way. People living in a world they experience as flat are bound to encounter some differences in comprehension and communication with those who interpret it as a sphere.

All thought takes place in a social context, but some ideas, like Darwin's, have immeasurably important social consequence, so that the prospect of radical change may have the effect of slowing the work of the thinker as he struggles to coalesce his ideas and to raise them to a new level. Darwin knew his theories were in irreconcilable conflict with conventional thought and would be denounced by ecclesiastical and other authorities and resisted by fellow scientists. In tracing the development of Darwin's thought, Gruber shows the retarding and potentially crippling effects of fear on the unorthodox thinker.

REFERENCE

Piaget, J. *Science of education and the psychology of the child*. New York: Orion Press, 1970; Paris: Denoel, 1969.

CHAPTER 2

FROM PREOPERATIONAL TO CONCRETE THINKING AND PARALLEL DEVELOPMENT OF SYMBOLIZATION

Hermina Sinclair

As yet in only a few countries in the world do some educators feel that Piaget's theory of cognitive development can help to bring about profound improvements in the educational system. In even fewer countries has an active beginning been made in this direction, and wherever this has happened, it has been the work of a very small number of dedicated people.

It may seem surprising that a theory that explicitly states that its aim is the study of the modifications of human knowledge through the ages, and of the development of knowledge in the child, has had comparatively little impact on education. However, as soon as one becomes a little more familiar with the problems Piaget has studied, it becomes clear that an application of his work to education is far from obvious. In all his experimental procedures, and in all the problems he sets, his subjects deal with fundamental concepts such as classes, relationships, time, causality, and space. His concerns are the development of knowledge, not skills or information.

Consider the question "When did World War I start?" This particular information can be found in history books. But what about the question, "Mummy, when is 1914?" asked by a

6-year-old child in a country where history is taught at a remarkably early age? A difficult and sensible question indeed for a youngster who, as Piaget has shown, still has a very peculiar concept of time. Such a child can think that as the years go by, he may catch up in age with a brother 5 years his senior; he can think that his watch does not measure time in the same way when he runs as it does when he walks slowly. Being able to recite the names of the months in the right order, even being able to count to 20, will not help a child of 6 to understand historical dates.

The child similarly interprets many other bits of what is commonly called "knowledge": It is very easy to look a fact up in a book, it is even fairly easy to memorize it, but to teach the underlying framework that alone will give meaning to the information is a totally different matter. One can teach a child how to measure distances with a ruler and how to calculate an area. But many children are taught this skill prematurely—that is, at the stage of thinking when, if in their presence you cut a sheet of paper into thin strips, glue them all together into a long ribbon, and ask the children if the surface is still the same, they will answer "of course not." For these children the necessary framework of real knowledge is absent, and the acquired skill probably will soon be forgotten as well.

Information and skills are not what Piaget's theory is about; however, they are what school is traditionally about, and they have usually been taught as if the framework provided by the real knowledge is automatically present in the child (at least from the day he starts going to school). When Piaget says that children are not miniature adults he does not mean that children want to run and play in the mud and that they are careless in crossing the street; he means (and he has amply shown that this is so) that children do not think like adults and that their fundamental knowledge is differently structured—not that it is simply unstructured due to lack of information and skills. Our IQ tests reinforce this idea. We all know that as soon as a child's IQ is measurable it remains stable more or less for the rest of his life. However, IQ tests measure the child's skills and information against what is supposed to be the norm of his age group, whereas Piaget's experiments show us how the child's fundamental knowledge that governs the way he thinks changes from age group to age group.

Society at large must change its ideas concerning the real aims of education before full profit can be derived from a theory such as Piaget's. Nevertheless, even now, when educators are bound by programs and examinations, much can be done to give meaning to school-learning and to avoid training in skills for their own sake and handing out information that cannot be assimilated. Thus it is vital to know something about how the fundamental knowledge is acquired and how thinking functions in the young child, if only to be able to fulfill the exigencies of the educational set-up as it is today.

Another point that immediately becomes clear when one reads Piaget is the fact that the way in which knowledge is acquired is not through the senses from the outside sources, but through action upon the environment and interaction with the environment. This does not mean that the child should spend his early school years digging in sandpits and making mudpies, progress to constructing buildings out of bricks, and then make up systems of pulleys and levers, but it does mean that looking and listening—traditionally and even at the present time the main ways of getting information—are not good enough.

Some of our recent studies are instructive about what is involved in the drastic change in children's thinking between the age of 4 and 8. Others suggest what happens when the child is presented with the two main forms of representation of reality—words and pictures—when his basic way of approaching reality is different from that of the adult's.

Hermina Sinclair

Cognitive Development between Four and Eight Years of Age

The most important characteristic of the changes that take place during the years between 4 and 8 is the following: Around the age of 6 or 7 (with differences among individual children and differences according to the context for one and the same child) a child begins to realize that the actions he performs on objects can be mentally canceled out. For instance, if he has put a big toy truck into a deep drawer that contains other big toys such as dolls and teddy bears, this constitution of an elementary class or collection of big objects will now no longer stop him from understanding that the truck can at the same time belong to a collection of toy cars. The action of including it in the collection of big toys can be mentally canceled out, and the same object can be put with the collection of motor vehicles. Children also discover relations of reciprocity; for instance, they begin to understand that if one stick is bigger than the other, the second is shorter than the first. Reciprocity and annulment are two different aspects of what Piaget calls "reversibility," a fundamental notion that gradually takes root in children's thinking. At first it is applicable to actions performed on discontinuous elements and later to the transformation of continuous quantities. Reversibility implies the construction of a coherent system of operations that, unlike the actions of the earlier period, can be effected mentally and that, instead of contradicting one another or simply being juxtaposed, now reinforce and sustain each other.

In the well-known experiment of pouring liquid from one of a pair of identical glasses into a narrower glass, the 6-year-old child knows very well that nothing has been added or taken away; but he finds it impossible to reconcile this fact with the reality of the higher level of the liquid in the narrower glass. Though he can, of course, perform the inverse action of pouring the liquid back again, he cannot perform this action mentally; he does not understand that the initial and final states are linked by a transformation that can be totally (not only approximately) canceled out by the inverse transformation. This link provides the insight into the reciprocal relationship: Now the liquid goes up higher, but it is in a thinner column; in the original glass it was lower, but the column was broader.

The conservation experiments show us, especially through the child's explanations and reactions to countersuggestions, that a grouplike structure containing an identity element is in the process of being built up. The existence of an identity element is in fact the better known implication of the acquisition of conservation; but it must not be forgotten that this is indissociable from the existence of the operational structure. In certain experimental procedures it is the appearance of quantitative constants that is the more obvious—so much so that the experiment with the balls of clay has led to misconceptions. Reversibility does not mean understanding that balls can be made into sausages and sausages into balls. In other experiments—for instance, in classificatory problems—it is the construction of operations that is the more obvious. That is to say, neither the identity element nor the operational structure can exist separately.

Though Piaget has formalized this first coherent framework of thinking in logical terms, it is not only in the logico-mathematical field that reversibility reveals its importance. The same system of operations also opens the way for the child's first observation of consistency in the way objects behave, in other words, for the first ideas of law and order in the universe. In his social interaction, in his understanding of words such as "foreigner" or even "brother," a very similar change takes place. At this stage of development children say that rules of a game have to be kept constant, "otherwise you wouldn't know where you are"; but a smaller player's size can be compensated for by allowing him a shorter distance (to shoot his marbles, for instance) or by letting him start a little earlier. Only now with reversibility does it become possible to take another person's point of view. Piaget tells the story of the little boy who announced that he had two brothers, Peter and Tony; however, when asked if Peter also had two brothers, he said "No, he's got one, Tony." At his age this little boy could see himself only as someone who *has* brothers, not as someone who *is* a brother. In his verbal expressions, the 6 to 7 year old also shows many symptoms of the same change, which is essentially the acquisition of reversibility. When shown a toy truck that pushes a toy car, the 5 year old will say that the truck pushes the car; but when asked to start his sentence with the word "the car" he may announce that this is impossible: "If you start with the car, it's the car that pushes, and you did it the other way round." From 6 on the child says the car is being pushed by the truck (Sinclair and Ferrero, 1971).

Reversibility can thus be considered a very general explanatory principle for the changes that take place between the ages of 5 and 8. This does not mean, of course, that it explains everything; no doubt many other factors intervene in different degrees in fields other than the development of intelligence.

However, reversibility by itself does not explain how the change comes about. One of the tenets of Piaget's theory is that every novelty is a result of a progressive construction, in which self-regulatory mechanisms are the explanatory factor. Similarly, in biology it is no longer possible to draw a dividing line between the organism with its innate structure and the influence of the environment in the child's development. Novel behavior is seen as resulting from a recombination of already existing mechanisms. Thus, if the profound change that takes place during the years from 4 to 8 can be characterized by the novelty of reversibility, then we must ask how earlier acquisitions prepare the way for this new development.

At the end of the preverbal sensori-motor period, the child acquires the permanency of objects and the structuration of their displacements in space at the level of pure action. This prefigures the operational structure that appears around the age of 7. But between these two stages there is the preoperational period during which the change is gradually prepared. This period has been negatively characterized by a lack of reversibility, a lack of decentration, and the absence of stable, quantitative constants. In some recent publications, however, Piaget has started to specify its positive characteristics. At the moment, in order to get a deeper insight into this period, work is in progress in Geneva with children from 3 to 5 years old. It seems useful to consider separately the two aspects that should not be dissociated—constants and action- or operation-structures. At every stage in development these two aspects are no more than two sides of the same coin; nevertheless, certain types of behavior clarify one rather than the other.

ACQUISITION OF CONSTANTS

Starting off with object-permanency and the group of displacements, we can consider that object-permanency is the first cognitive constant that, at an operational level, evolves into the construction of quantitative constants (as they are seen in the conservation experiments). The grouplike structure of the coordinations of displacements in space develops into the operational structure of reversible, interiorized actions. If we first take the development of the cognitive constants, we can very briefly sketch the following evolution. In the first cognitive constant, object-permanence, the subject's actions and the object itself are inextricably linked. Psychologically speaking, object-permanency means above all that the object has now become "retraceable." A little later, objects acquire a *qualitative identity* that is no longer uniquely a function of the act of searching for them, but is due to the fact that the child realizes that several actions can be performed on the same object without changing its basic identity. For instance, a piece of wire can be twisted into the shape of a pair of glasses, or scissors, but these different shapes are linked by the fact that it is the same wire that was used to produce them. Though the child will put glasses on and pretend to read the newspaper, he knows, and he will say so if asked, that it is the same piece of wire that earlier was used for something else.

The next step is taken when the child begins to make a distinction between the permanent and the nonpermanent qualities of objects. Color, suppleness, and material of the wire are permanent, but its shape is not. For children below the age of 7 or 8, certain changes of form imply a change of length. Nevertheless, the identity of objects has become more objective, in the sense that it is now based on their permanent qualities rather than the actions the subject can perform upon them.

The great novelty of the operational period is the change from qualitative identities toward quantitative constants. The first one of these is numerical conservation, which manifests itself in the fact that in answer to the well-known questions about the numerical equivalence of two collections of objects of which one row is spread out to go beyond the limits of the other (red and blue counters, for instance), the child no longer thinks that there are more red than blue counters. He now affirms that there are still enough red ones to cover all the blue ones, that no blue one will be left over, and that he will not need any extra red ones if he wants to try it out.

STRUCTURE OF ACTIONS

If we consider the second aspect, that of the structure of actions or operations, we see that very early the child has a grasp on what may be symbolized as one-way mappings and what are, psychologically speaking, the functional dependencies. For instance, he knows that the movement of a block depends on the push he gives it; the harder the push, the further the block will go. This dependency is at the same time both a real, physical dependency and a conceptual dependency. He knows that a push (y) makes the block move (x), whether he himself does it, somebody else does it, or another object hits the block: $x = f(y)$. The move (x) depends on push (y); and knowledge of (x), the movement, depends on knowledge of (y), the quality of the push.

These dependencies constitute a kind of semilogic, and their one-way character has been demonstrated in several studies. For instance, in an experiment by Grize et al. (1966), the child is shown a toy truck that picks up counters in front of a number of dolls. The color of the counters corresponds to the color of the doll's dress. The dolls are arranged on the table in a fixed pattern, but in neither a straight line nor a circle. The first questions concern the arrangement of the counters inside the truck. Which will be first? Which will be last? Why is the red one next to the yellow one? The 4-year-old child understands and explains that the order of the counters in the truck is dependent upon the itinerary of the truck: If it goes to the blue doll first, the blue counter is first, if it goes to the yellow one next, the yellow counter will be next to the blue counter, and so forth. But, surprisingly, if he is asked to reconstruct the itinerary of the truck, he is incapable of doing so; the mapping is only one way, and he does not understand that the order of the counters in the truck determines in turn the itinerary of the truck, just as the itinerary determines the order of the counters.

However, even this incomplete semilogic is an important development and a necessary stage that the child has to complete before he can acquire reversibility. The well-known experiment with the balls of clay can be reformulated in terms of functional dependencies. At first, a dependency is established between actions and their effects: If one rolls (x) the clay, it becomes longer (y): $y = f_1(x)$. Another function is then established: If one rolls (x) the clay, it becomes thinner (y'): $y' = f_2(x)$. Both dependencies combine to $(y,y') = f_3(x)$. That is to say, if one rolls the clay it gets longer *and* thinner. Finally this covariation between y and y' is directly expressed without the necessity of linking it to the action itself, and a reversible function is obtained whereby getting longer is exactly compensated for by getting thinner and vice versa: $y = f(y')$ and $y' = f(y)$.

Development of the Symbolization Process

The second part of this chapter concerns the development of the symbolization processes. How are these influenced by the cognitive development? Unfortunately, far too little is known about the way representation and symbolization develop. The acquisition of language between the ages we are concerned with is almost totally a mystery. However, Genevan research has brought to light a number of striking examples of the ways in which children deform information which is presented in a visual or verbal manner. If nothing else, these examples provide a serious argument against the passive look-and-listen methods of teaching.

In the following section are some examples of children's behavior in memory tasks of different kinds (Piaget and Inhelder, 1968): memory of an action performed by the experimenter in front of the child, memory of a drawing presented to the child, and memory of sentences spoken by the experimenter. In all cases, it was possible to prove that the deformations were not due to a too heavy memory load in the general sense. For instance, for sentences it is quite simple to devise control sentences that are as long or even longer than the sentences proposed and that the children can reproduce without any difficulty. In all cases, deformations occurred because the children simply did not apprehend the action, the drawing, or the sentence in the way an adult would, and their own productions were most revealing as regards their thought processes. The ensuing examples were chosen to touch upon different fields of knowledge: simple physical events, logical problems, and linguistic structures.

In a first experiment a problem of transitivity was involved: If there is more liquid in glass *B* than in glass *A*, and more liquid in *C* than in *B*, is there more liquid in *C* than in *A*? However, it was not the logical problem (difficult to solve and difficult to remember even after the age of 7) that was relevant, but a curious phenomenon that showed how even a simple action like the pouring of liquid from one glass to another can be deformed in memory. The experiment used 2 glasses, 1 with red liquid (*A*), 1 with yellow (*B*), and 2 beakers (*C* and *D*) with no liquid. The liquids were poured into the empty beakers. Then the yellow liquid was poured into the glass that had contained the red liquid and vice versa, so that at the end the contents of two differently shaped vessels were reversed (see Figure 2–1). When the children were asked to tell us what they had seen, the 4 and 5 year olds maintained that we had poured the yellow liquid into the glass with the red liquid and vice versa. We wondered whether this was a kind of abbreviated description of what really had happened, and we showed them the 4 glasses with the liquids in the original position. To our surprise, the children actually took up the 2 glasses that were filled with liquid and tried to pour simultaneously the yellow liquid into the glass with the red liquid, and the red liquid into the glass with the yellow liquid. Questioned whether they really thought that this could be done, they maintained their answer: "Yes, if you are clever enough." "Won't the yellow and the red get all mixed up?" was our next question. Many hesitated or simply said "No." One child said "Yes, maybe, but it will unmix itself in the end."

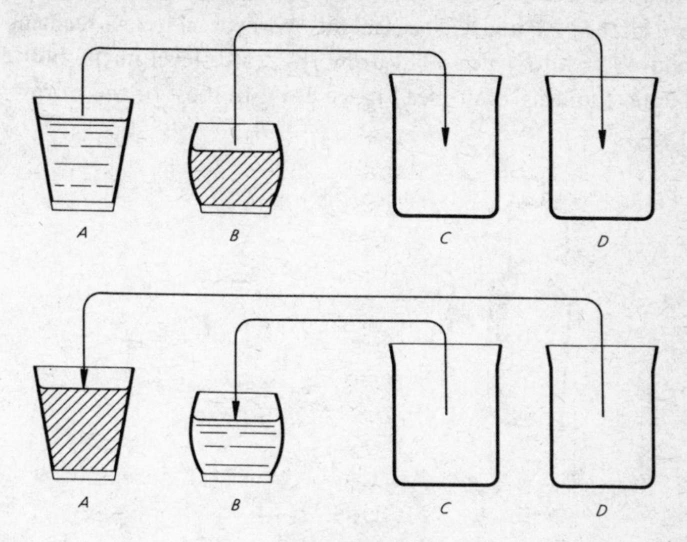

FIGURE 2–1
Transitivity Problem

Another example of a deformation of physical reality in a quite fantastic way was the children's memory drawings rendered after they had been shown a picture of a bottle with liquid in it lying on its side next to a little toy car (see Figure

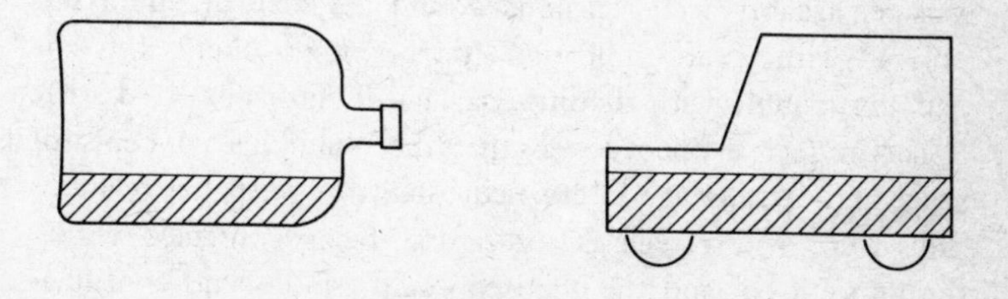

FIGURE 2–2
Memory Task Using Drawing

2–2). To reinforce the horizontality of the liquid level the liquid was colored red and the toy car had a red horizontal strip across its body. Despite this perceptual reinforcement, children of 4 to 5 years old drew the water level in the bottle in a most unusual way (see Figure 2–3). In most of their drawings, the bottle was drawn as being upright instead of lying on its side. This is not surprising, since bottles are, after all, more often seen in this position than seen lying down. However, the liquid inside the bottle hung along one of the long sides in a kind of vertical strip. The children could unquestionably draw a bottle and its contents; in many drawings the bottle was remarkably well reproduced, and the strip of liquid was colored with great application. Moreover, no one at this age had any trouble with the little car and its horizontal red strip. What in fact happened was that the children's concepts of space and of causality intervened. At their age spatial relationships were mainly seen as intrafigural, being contained within the figure itself, and the children could establish no coordination with outside points of reference. The bottle was put in its usual position; but the relationship between the liquid and the

FIGURE 2–3
Child's Solution to Memory Task

long side of the bottle, inside the one figure as a contiguity relation, was kept constant and gave rise to the representation of an impossible physical event.

Another example concerns a logical, numerical problem. The children were shown an arrangement of four matches in a straight line, underneath which there were four matches in a zigzag line, so that the extremities of the straight line went well beyond those of the second (see Figure 2–4). The chil-

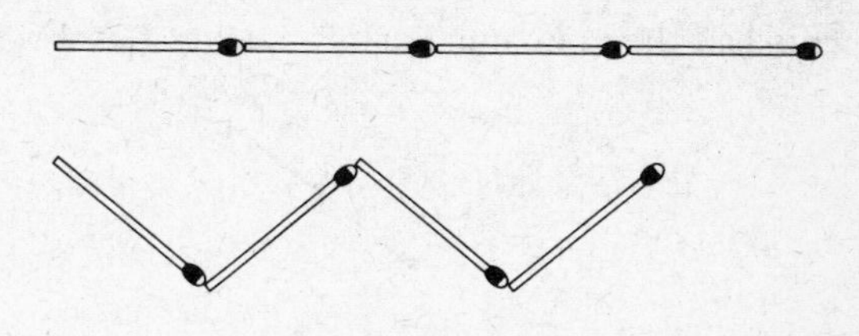

FIGURE 2–4
Logical Numerical Problem

dren's attention was drawn to the fact that for every match in the top line there was a match underneath, or that, for the older ones, there were 4 matches above and 4 matches below. In one example of what the 4 to 5 year olds did, by way of deforming reproductions, two types of drawings were especially interesting. The first correctly represented the 4 matches in the upper line, but the lower line was composed of 6 or even 7 or 8 matches in a zigzag, carefully drawn so that the extremities of the bottom arrangement coincided with those of the top straight line. The second type again correctly represented the 4 matches in the top line, and the bottom one again had 4 matches, which as in the model were arranged in a kind of *W*; but the zigzag matches were drawn so much bigger than the straight-line model that again the extremities of the arrangement coincided with the upper line (see Figure

2–5). To check whether this representation was typical of the memory of discrete objects arranged in a certain way, we also gave the children a model of two pieces of wire, one straight and one twisted into a *W*. This time, no difficulties were encountered and drawings were more or less correct even at this early age. The deformed reproductions of the match arrangement resulted from the typical difficulties of the preoperational child in judging numerical equality. The children seemed to have accepted the fact that in the model both arrangements had the same number of matches (4). From this,

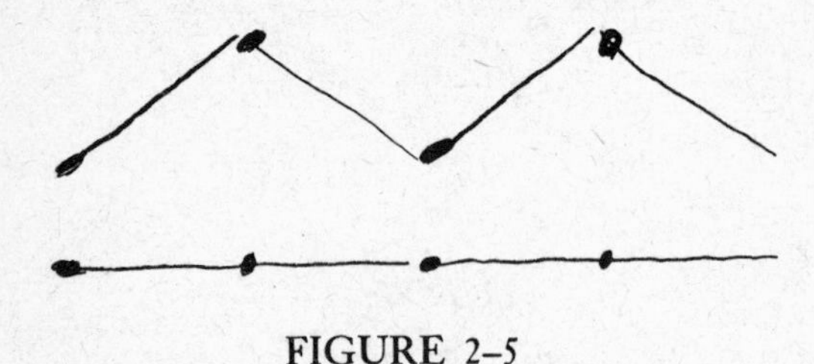

FIGURE 2–5

Child's Solution to Logical Numerical Problem

they inferred that the lines should start and end at the same point. In fact, preoperational children judge numerical quantity by the space occupied; a line that "goes further" has "more" objects. Therefore, they drew two arrangements with coinciding extremities, thinking that such arrangements had the same number of elements. As a result, they had to draw many more matches in the zigzag arrangement than in the straight line! The other solution, whereby the matches themselves were drawn at twice their length, was at a rather higher level. In that case numerical equality was kept constant, at the price of sacrificing the length of each individual match.

In many of our language experiments, the children were asked to repeat sentences from memory; as in the other examples described, the repetition was often requested immediately after the model sentence. To see whether the child understood the particular type of sentence, he was asked to act out what had been said, either before or after he had repeated it. Depending on the age of the subjects and the type of sentence presented many deformations were observed. For instance, when told: "The car is pushed by the truck," 4 year olds and some 5 year olds repeated, "The car pushes the truck," just as, when asked to act out the sentence, they also reversed the role of pusher and pushed (Sinclair and Ferrero, 1971).

Relative clauses continued to be deformed well beyond the age of 4 or 5. For instance, a sentence such as "The cat licks the dog that pushes the monkey," was repeated as "The cat licks the dog and pushes the monkey," and was acted out in the same way.

These few examples of deformations of actions that the children have seen performed in front of them, of pictures and of sentences, have been taken out of their context, but they are sufficient to show how little we can rely on children to take in information presented to them in the way it is intended. On the contrary, in every one of the examples quoted, it is necessary to know something about a child's way of thinking in order to understand the deformations he introduces. Information coming from outside is assimilated to the basic knowledge the child possesses; if this knowledge is such that correct assimilation is impossible, then the information will be deformed.

Unfortunately, teachers, and especially teachers who have to deal with large classes, are forced to resort to verbal or pictorial representations. As every teacher knows, the answers to questions such as "What did I just ask you?" or "What did I just tell you?" are often extremely disappointing. However, this lapse is not, as is often thought, due to lack of attention or faulty memory; it occurs because the child does not simply copy what his eyes have seen or his ears have heard. He assimilates what he has seen or heard to his basic knowledge, unless, in the unhappy cases, he has already learned to mistrust his thinking and to rely on his copying capacity. Educators are generally agreed that this path should be avoided at all costs. It is not the ways in which information is transmitted that should be improved. No effort should be spared to find better ways of helping the child to develop his fundamental thinking abilities. And it is toward this goal that Piaget's theory can be of inestimable value.

REFERENCES

Grize, J. B., et al. *L'Epistemologie du temps*. Paris: Presses Universitaires de France, 1966.

Piaget, J., Inhelder, B., and Sinclair, H. *Memoire et intelligence*. Paris: Presses Universitaires de France, 1968.

Sinclair, H., and Ferrero, E. Etude génétique de la comprehension, production et répétition des phrases au mode passif. *Archives de psychologie*, 1971.

CHAPTER 3

RECENT PIAGETIAN RESEARCH IN LEARNING STUDIES

Hermina Sinclair

It is hardly necessary to explain in detail the general theoretical background to the Genevan learning experiments. However, a number of misconceptions as regards the Genevan position on learning are so widespread that a few words of background will serve as a useful introduction.

A first group of learning experiments was undertaken in the early 1960s by the Centre d'Epistemologie in Geneva.[1] As is often forgotten, these experiments were undertaken with a very specific aim: to see whether the providing of information from the outside—that is to say, giving the child the opportunity for verification of the predicted outcome of an action—could change the child's reasoning. In these procedures strictly empirical epistemological tenets were to be applied to the learning of cognitive structures. Almost universally the results were negative. However, Piaget and his collaborators did not conclude that any kind of learning procedure would be useless. To the contrary, such a conclusion would go counter to the developmental theory that is resolutely interactionist and constructivist. The results meant only that empirical methods, whereby the subject has to accept a link between events because this link is imposed upon him, do not result in progress; progress results when the subject himself discovers the link. This active discovery of links is what happens in development; it is therefore called spontaneous—maybe unfortunately

—for development is always the result of interaction. It remains true that the subject himself is the mainspring of his development, in that it is his own activity on environment or his own active reactions to environmental action that make progress. Learning is dependent on development, not only in the sense that certain things can be learned only at certain levels of development, but also in the sense that in learning—that is, in situations specifically constructed so that the subject has active encounters with environment—the same mechanisms as in development are at work to make for progress, and, if there is progress, the same structures result.

The fact that in different cultures and subcultures the same line of development is found although acquisition ages vary according to whether or not the child is exposed to a stimulating social environment shows that environment can accelerate or retard development, but only rarely can it change its course. Bovet (1970) has found some cases where a certain deviation, as she calls it, from the developmental direction is present; but such deviations are only temporary and rejoin the already known stages and substages. Interestingly, this rejoining is present both when different age groups are interviewed cross-sectionally and when the one age group is given a certain number of training sessions.

Therefore, when Inhelder, Bovet, and the writer started learning experiments, it was not at all in order to show that no progress could be obtained; on the contrary, we hoped that methods based on what was known of development *could* have an accelerating effect in contrast with methods that were based on empirical tenets. On the other hand, it was never our intention to find the most appropriate methods of training (even supposing that they could have been determined in a general way), but rather to learn more about the mechanisms by which transition from one substage to another takes place, from a psychological point of view. Such functional mechanisms have been described by Piaget as regulatory mechanisms, resulting in equilibration; or as adaptation, which is seen as a coordination between assimilation and accommodation. Contrary to popular views, the structures that Piaget supposes underlie the behavior observed when the child is asked to solve certain problems do not in themselves provide a description of the way the transition takes place. To take a well-known example, conservation experiments have revealed the arguments children use to justify their correct answers. Piaget has shown that a certain formalization in terms of grouplike structures can account for a great number of observable behaviors (constructions children make or arguments they give) within a certain stage of development. But one cannot take the arguments one by one and try to teach them to the children, nor take "an operation" (for example, the inverse operation) and expect that the child will "interiorize" this operation by making him transform and then bring back to the original shape balls of clay or lengths of wire. This application is a distortion of the theory, in which observable behavior and inobservable structures are never mixed and in addition structures and mechanisms of transitions are clearly distinct. Moreover, because the main characteristic of mental operations is that they always form a grouplike structure, to talk of one operation is in a sense a contradiction in terms. Though from the point of view of logic the formalization given to the underlying structures may not be the only possible one or the most adequate one, this characteristic remains fundamental.

Our thoughts when we started the program of learning experiments ran as follows. Given the fact that children's thinking develops through encounters with environment (in its largest sense) and that new behavior results from a recombination of already existing action or operation schemes, we might be able to arrange the encounters of children with environment in an ordered and accelerated way. If we thereby succeed in obtaining progress we might be able to observe during the sessions the successive moments when this supposed recombination takes place. Accordingly, we used what is already known of the different obstacles to the solution of our various problems to build our learning procedures. Often, however, our knowledge of these obstacles was general and theoretical rather than precise and pragmatic, and we had to do many preliminary experiments to get a better insight into the children's difficulties. Other more specific problems we wanted to investigate were the reason for the so-called *décalages*, or time lags (particularly between the different conservations), and the links between different fields of knowledge, for instance, between geometrical and logical concepts. Only the first problem is touched on here: Is it possible to observe changes in behavior during learning sessions that clarify the mechanisms of gradual adjustment (what Piaget calls "regulations") and recombination of existing schemes? We think it is.

I shall give some examples of such behavior in three different learning procedures: (1) one concerning a problem of conservation of length in the field of geometry; (2) one concerning a problem of class inclusion in the field of logic; and (3) one concerning a problem of conservation of volume, which is at least partly a problem in physics. The learning procedures for the conservation of length and of volume were devised by Bovet (Inhelder and Sinclair, 1969). Most of the procedures gave excellent results in terms of the acquisition of the concepts involved. However, in the context of this chapter quantitative results, pre- and post-tests, and learning curves will not be mentioned. The examples given are meant to illustrate two fundamental problems: What are the difficulties children encounter in the acquisition of a certain concept, and what is the nature of the transition to higher order reasoning patterns, when such transitions take place during the training sessions?

Conservation of Length

In cross-sectional experiments a *décalage* of about two years was found between the acquisition of elementary numerical conservations (that is, the understanding that different spatial arrangements of the same number of discrete objects do not change the numerical extension of the collection) and the acquisition of the conservation of length. Several situations were devised to make the children realize that they could use the way they solved problems of separate elements when faced with problems of continuous lengths. For instance, matches were glued onto tiny toy houses so that roads could be built with different contours, whose lengths could be evaluated by the number of houses along them. We wondered whether children who in a pretest had no trouble understanding that a change in the disposition of 1 of 2 lines of houses (without matches) originally set up in an optical one-to-one correspondence did not change the numerical extension of the 2 collections, would immediately understand that 2 roads made by the matches glued to these houses would also remain the same length. In this situation it was easy to ask questions alternatively on the number of houses and on the length of the road: e.g., if you walk along here, do you encounter as many houses as on the other road? Or are there more houses on the one road than on the other? If you walk along here, do you have further to go than if you walk there? Will you be just as tired, less tired? For some children there was no connection between the two types of questions. The number of houses? It was the same. And the roads? They were different—one was much longer, because it went further (the straight road, compared to a zigzag road); you'd be more tired, because you had farther to go. And so forth. Other children seemed to catch on and argued: Same number of houses means same number of matches; same number of matches means same road.

However, in a second part of the experiment children were asked to judge the comparative lengths of road that the experimenter had constructed using matches of either equal or unequal length. In addition they were also asked to construct roads themselves, following different contours and starting at different points from the experimenter's, with matches of different length.

In fact, it is no proof of the acquisition of the concept of conservation of length or of the capacity to measure continuous length if one can solve problems where lengths are constructed out of elements of equal size; measurement, as Piaget has shown, involves the capacity to partition a continuous quantity and the understanding that units have to be used, which themselves have a constant length. Using matches of equal length means that the experimenter has already solved part of the problem for the child, who can now simply discard his intuitive solution (whereby he judges distance by points of departure and points of arrival) in favor of a counting procedure, where he judges by the number of elements.

In this series of problem situations a number of behaviors on the part of the children should be discussed here. Perplexities, contradictions, questions, and compromise solutions in construction problems all seem to fall into the category of behavior that can clarify the problems of the actual mechanisms of transition.

Having "learned," it seems, that length of a road can be judged by counting the number of matches, and having correctly solved a number of problems dealing with matches of equal length, one child—Cath—is faced with the following situation: 7 shorter matches make a road of equal length to that of 6 longer matches, the two roads being in a straight line and directly parallel to each other (see Figure 3–1). This

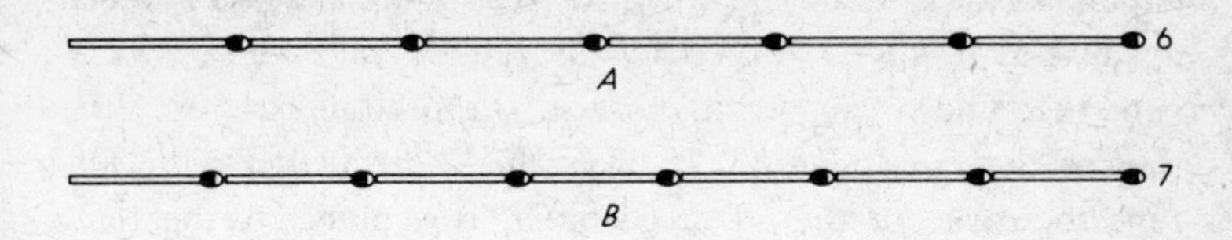

FIGURE 3–1
Conservation of Length Problem (Cath)

situation poses no problem for children who do not have any conservation of length; they correctly judge the roads to be equal. Cath, however, now announces that *A* has less far to go than *B*, for there are 6 matches as against 7. She explicitly refers to *A* as being less tired, and to *B*'s road as being longer. When discussing the situation with the experimenter, she changes her opinion several times: "same length, because I can see it, they go just as far"; and "not the same length, I count the matches, it makes 6 here and 7 there"; but at no point in the discussion does she refer to what would conciliate these two different answers, that is, the unequal length of the matches. When asked about the following situation—4 matches to form one road in a straight line, 6 matches to form a parallel road in a zigzag pattern, departure and arrival coin-

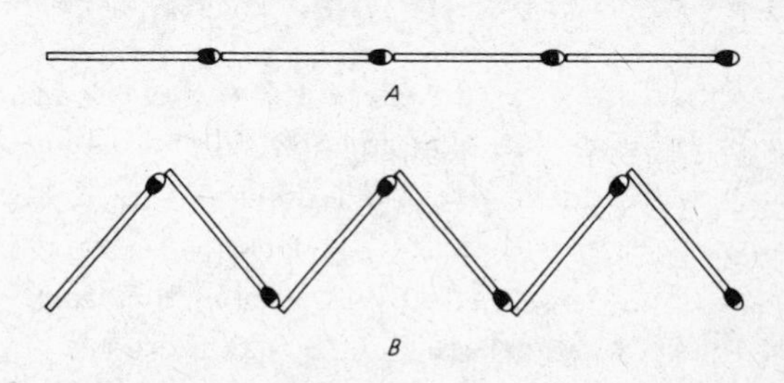

FIGURE 3–2
Conservation of Length Problem (François)

ciding, all matches of equal length (see Figure 3–2)—François answers as follows: "The roads are exactly the same . . . except that you've put a bit more in the bottom road so that they're the same length." It takes the experimenter a bit of time to work out this involved bit of reasoning. During that time François himself seems to go over his reasoning again and becomes more perplexed. Fortunately for us, he is ready to tell us about his problems and says: "But then . . . why *are* they the same? That's what I'm wondering about . . ."

A third example of conservation of length is taken from the following experiment. Georges is asked to construct a road of equal length to a zigzag road, but in a straight line, using same-sized matches (see Figure 3–3). He counts the matches in the

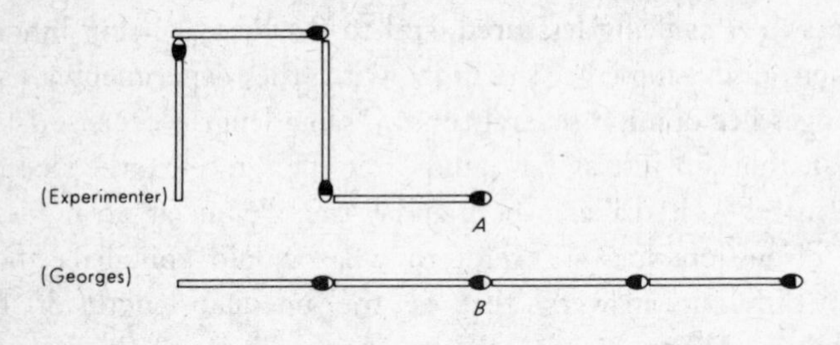

FIGURE 3–3

First Conservation of Length Problem (Georges)

zigzag road, finds that there are 4 and puts down 4 matches in a straight line. When the experimenter tells him that another child in this situation says that *B* is longer than *A* because it goes further, he scorns this idea: "That's just funny, there's 4 matches and 4 matches and he thinks it's not the same!" However, in the next situation (see Figure 3–4) there are 4 matches in a straight line as against 5 in a zigzag, but coinciding extremities. When Georges is asked whether the roads are the same length, he changes his mind and says that they're the same. He correctly counts 5 matches as against 4, but "no, the zigzag road is not longer; it's just the same; it goes just as far." We go back to the situation where he has correctly constructed a road of the same length. This time he does not simply count 4 in the model road and then put down 4 in his own road; instead, while he builds his road he touches with

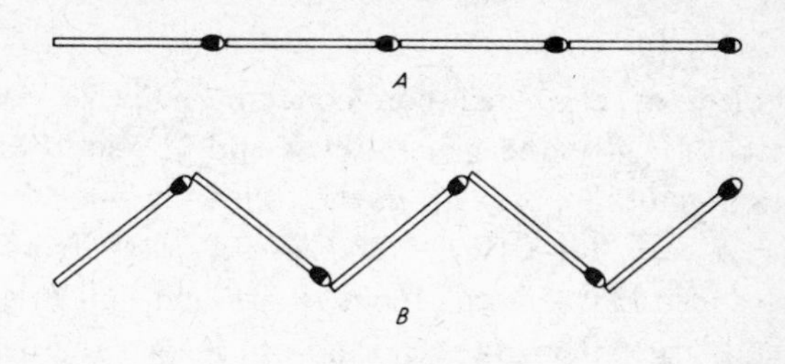

FIGURE 3-4

Second Conservation of Length Problem (Georges)

his other hand each of the matches in the model road. When we go back to the other situation, he applies the same method, removes one of the matches in the zigzag road so that the extremities no longer coincide, and judges that now they are the same length, but not before. When we remind him of what he has said earlier, he explains: "because I didn't count properly, because that came to the same place" (showing the extremities). This answer, too, takes a bit of working out; in fact he *had* counted correctly, 5 as against 4; but he had taken no account of this numerical inequality, discarding it in favor of a judgment only on "going to the same point and starting from the same point." In other words, he had counted correctly, but he had not been able to make the correct use of his counting, because his immediate reasoning was bound to the topological configuration.

When we look at the results in our post-tests, which present the conservation of length problem in the classical way, using a length of wire which is either displaced or twisted, we see that children who show the types of behavior just described make progress, and often succeed completely.

Quantification of Class Inclusion

The problem of class inclusion concerns additive compositions, that is: If B is the general class and A and A' are the subclasses together making up B, the following operations obtain: $A + A' = B$, $B - A' = A$, and $B - A = A'$. From this it can be deduced that if both A and A' are non-null B is larger than A and larger than A'. According to Piaget, the inclusion of class A in class B provides the relationship that proves the statements "all *A*s are some *B*s and there are more *B*s than *A*s." Certain subjects can agree with the first statement even though they do not understand the second. Complete understanding of the concept of inclusion implies the understanding of the link between the operation $A + A' = B$ and the operation $B - A = A'$.

It is only at the level of concrete operations (by definition, reversible and interiorized) that the child becomes capable of working simultaneously in a general class defined by a more embracing criterion (e.g., flowers), and on subclasses with a stricter criterion (e.g., roses). At the preoperatory level the child does not conserve the whole when he has to compare it to one of the parts; his mistake is that, when he begins by mentally evaluating A, he isolates it from the whole B and can only compare it with A' and not with B. When faced with a bunch of flowers containing a great many roses and a few tulips, and asked if there are more roses or more flowers, the child replies that there are more roses; if he is then asked "more than what?" he often answers "than tulips." Except for a child's hesitations or corrections, the original form of the test does not permit us to determine substages in the acquisition. Moreover, it relies uniquely on the child's verbal answers and lacks situations where the child himself has to construct classes and subclasses. The theoretical analysis seems to indicate, as did the childrens' answers, that the main difficulty lies in the fact that the child is asked to compare within one collection the extension of a subclass with that of the total class.

With these considerations in mind, we constructed a learning procedure in which the children had to construct by themselves collections within which the subclasses varied in numerical extension while the extension of the general class was kept constant. Secondly, they were asked to compare a subclass in one collection with the general class in another, numerically equal collection, before having to perform this comparison within one collection only. We hoped that in the first part the child's manipulations (the adding or taking away of elements, the progressive corrections) might reveal the progressive adjustments in his actions before these gave way to the immediate correct answer made possible by the operational system. We also hoped that the answers to the questions on 1 and on 2 collections might again reveal perplexities. Below is a brief outline of the technique to make the examples more comprehensible.

Experimenter gives to girl doll: 6 *F*s (pieces of fruit), for example, *AAAAPP* (4 apples and 2 pears). The child is asked to give the boy doll "Just as many pieces of fruit,[2] so that both dolls have just as much to eat, so that nobody is jealous, but the boy doll has more pears because he likes them better than apples." The instruction is repeated as often as necessary in different forms. The situations can be varied, made easier or more difficult, by taking more or less items as the model collection. The reader should keep in mind the *E*'s (experimenter's) model here as being *AAAAPP* and the instruction is that the child should give to the boy doll "the same number of pieces of fruit, but more pears."

Following are the types of behavior observed, from the lowest to the highest level (from I to V).

Ia and Ib. "It can't be done . . . well, I don't know, maybe it can, sometimes . . . but I can't do it, one is going to be jealous, or it isn't right . . ."

II. The child gives the boy doll *PPPPAAAA*. "I was thinking, the apples are the fruit, so I gave 4, 4 apples, and then more pears, so I gave 4, that's more than the other one."

III. The child gives the boy doll *AAPPP*. "If he has to have more pears, but still the same to eat, then I have to give him less apples."

IV. The child gives *AAAAAA*—only apples—but the correct number.

Va. The child gives only one correct solution, keeping the numerical division constant: *PPPPAA* for *AAAAPP*. "One's got 4 pears and the other's got 4 apples, and one's got 2 apples and the other's got 2 pears."

Vb. The child gives several correct solutions.

The following excerpt is an example of perplexities that arise during the construction of collections. Here the reader should keep in mind the experimenter's model as being *PPPPAA* (4 pears and 2 apples), and her request is to give the boy doll the "same number of fruit but more apples." The following exchange takes place between the experimenter and the child.

E. "What have you given him?"
C. "Two apples and 4 pears."
E. "Can you remember what I've just said?"
C. "The boy is to have the same thing as the girl."
E. "The same thing as what?"
C. "The same thing of the other fruit."
E. "Yes, he wanted just as much fruit to eat as the girl, but he wanted to have more apples than her."
C. "We've got to add some then."
E. "Go on."

C. Child begins again. *AAAAAAPPPP*.
E. "Do you think that's all right?"
C. ". . . em . . . No."
E. Repeats instructions.
C. Child takes away all the fruit given the boy doll, ponders, and then seems to make a discovery. "We've got to give just apples then?" Child gives the boy *AAAAAA*.
E. "There we are. Now, what's he got?"
C. "Now he's got more apples than the girl."
E. "Right. Do the girl and boy both have the same number of pieces of fruit?"
C. "No, one's got more."
E. "Who?"
C. "The boy, I gave him 2 extra apples."
E. "You're sure? How many pieces of fruit has he got then?"
C. (Without looking) "He's got 8."
E. "Have a good look and count them carefully."
C. (Surprised) "He's got 6, too."
E. "Yes, you see, you did do the right thing."

Following are some perplexities during questions on 2 collections, i.e., *AAPPPP* to the boy doll and *AAAAPP* to the girl doll. The child himself constructed these collections, and it could have been supposed that he would have no difficulty in answering our questions. However, it is clear from the excerpt that constructing a collection by a step-by-step procedure is different from reasoning on the result.

E. "Who has more apples?"
C. "The girl."
E. "Who has more fruit?"
C. "The boy."
E. "Who has more pears?"
C. "The boy."
E. "Who has more fruit?"
C. "The girl . . . well . . . they both have more fruit."
E. "Really, how come?"
C. "If you look at the apples, then the girl has 2 pieces of fruit . . . 2 pears . . . then she has less . . . no, if you look at the pears . . . oh, no, they both have the same!"

The Conservation of Volume

The problem of conservation of volume involves first a series of situations concerning the well-known difficulty of dissociating weight and volume: Though bigger objects are often heavier than smaller objects, this is by no means a general rule; and secondly, though weight has something to do with the problem of the displacement of water when an object is immersed in it, it is not weight that counts, as young children usually think, but size. Thus the children were first presented with a number of toy barrels filled with metal filings (a bigger barrel could weigh less than a smaller one) and with scales to determine the respective weights of the barrels. They were then asked which one of 2 objects would make the water go up higher, in the case of: (1) 2 objects of the same weight, but different volume; (2) 2 objects of the same volume but different weight; and (3) 2 objects of different volume and different weight, in this case the bigger object weighing less than the smaller.

Typical compromise solutions were observed. If 2 objects have different sizes but the same weight, some children can cancel the latter factor and correctly judge according to size. However, if they have to anticipate what will happen if we immerse two objects of different weight but the same size, weight counts and they think that the heavier object is going to make the water rise higher. Curiously, when weight and volume vary inversely, weight can compensate for volume: "That one's bigger, but that one's heavier . . . it'll go up the same." After the child has seen the effects of the immersion, he says: "It changed its weight since the water has gone up higher." When he is asked, "Which one? Or both?" he replies "The heavy one." To the question, "How do you know?" he answers, "Because the bigger one makes the water go higher."

This child invents a reason (not at all absurd) to conciliate the two observations and supposes that certain objects change their weight when immersed in water. He is, however, forced to suppose that not all objects are subjected to this law, but at his level of development this does not bother him too much.

Other children, less advanced than this subject, simply refuse to accept the evidence of the experiments; they either maintain that the objects really weighed the same, that the scales "did not work right" or that "somebody pushed them," or they try to show us that, as they have predicted, the heavier object really makes the water go up higher. "You have to look like this to see it," they assert, bending their head to get an oblique view.

Further advanced children may demand a repetition of the experiments; they will then change their reasoning: "I thought weight mattered, but it doesn't. . . . it's just bigness that counts."

From our point of view, these perplexities and compromise solutions are extremely important indications of what mechanisms are at work during the transition. In all cases, the children go from one solution to another, applying one line of reasoning immediately after another. Both kinds of reasoning are incomplete and inadequate for the situation they are faced with. Efforts to conciliate the two different cognitive schemes result in some cases in a new construction that solves the problem satisfactorily. Sometimes, however, they result in the compromise solutions, in which the child seeks to compensate for one difference (for instance, weight) by means of another (for instance, size), although these differences involve what are for us irreconcilable parameters. In problems where the children get the opportunity of testing their predictions against reality, these perplexities may seem the outcome of the fact that the test results contradict their judgment. However, observations of reality are always interpreted by the subject (and interpreted in a different way according to his level of reasoning). The contradiction is therefore between two different interpretations of the same situation—in Piagetian terms, between two or more schemes. The real situation is only the occasion for a scheme to be applied. It is the experimenter's choice of situations that results in the appearance of different schemes, and it is an active recombination of schemes by the child himself that results in the acquisition of a new concept.

NOTES

1. These studies were supported by a research grant from the Ford Foundation (680-001g) and were completed with the help of a Fondy National Suisse de le recherchie scientifique (No. A 1336g).

2. The formulation of the questions is simpler in French than it is in English, because *un fruit* has a straightforward plural—*des fruits*, whereas in English a collective noun has to be used.

REFERENCES

Bovet, M. Piaget's theory of cognitive development, sociocultural differences and mental retardation. In *Social-cultural aspects of mental retardation (Proceedings of the Peabody NIMH Conference)*. New York: Appleton-Century-Crofts, 1970.

Inhelder, B., and Sinclair, H. Learning cognitive structures. In P. Mussen, J. Langer, and M. Covington (eds.), *Trends and issues in developmental psychology*. New York: Holt, Rinehart and Winston, 1969.

CHAPTER 4

COURAGE AND COGNITIVE GROWTH IN CHILDREN AND SCIENTISTS

Howard E. Gruber

Children and scientists are patently different in their thought processes, but not so different as it might seem. Piaget's work about the qualitative difference between the thinking of children and of adults is well known. But the important elements of similarity are rarely brought out.

An adult, like a child, may act on a preliminary set of premises until through his own experience he discovers their inadequacy. Even scientists like Charles Darwin, as we shall see, have this common quality of human thought.

Darwin is known and honored for his creative achievements. Children, too, have minds that are free, unfettered, and inventive, and their early years are filled with experimentation. These constructive human qualities are there to flourish except as they are impeded by repressive, external forces, the kind that had so much influence on the expression of Darwin's thought. By studying the mind of an adult scientist we may better understand the development of the child's intellect.

It seems reasonable, too, that a book on the educational applications of Piagetian theory should devote some of its attention to the history of science. Both the process of education and the process of science exhibit the creative tension between the structure of knowledge as it exists at a given moment and the human organism's struggle to adapt its own knowledge to changing circumstances. Although the fields of education and of child development have many advantages for those who would study the growth of thought, the process of scientific inquiry as an object of study has at least one of its own. When we link the subjects of education and child development, we are prone to make an error that represents a kind of "magical thinking": We wish the child to grow up and in fact he does; we therefore attribute his growth to our desires and our efforts (Piaget, 1930). This questionable causal attribution provides the main justification for adult efforts to educate children.

In recent years we have become increasingly aware that adults do not teach children some of the most fundamental ideas; at best, we help to provide circumstances in which children discover what they must know. Before Piaget's work no one ever dreamed of teaching such elementary concepts as the conservation of matter; yet even in those ancient days children universally developed those concepts. Now that Piaget has elucidated this feature of cognitive growth, many educators are prone to incorporate his findings into the professional structure of which they are masters: Teachers teach and children learn; therefore let us teach conservation.

This kind of magical thinking is hardly possible in studying the process of creative scientific enquiry. In principle, it seems, no one else can teach the scientist; he must discover whatever he must know. If we can glimpse some common features in these two processes of discovery—in children and in scientists—we may avoid the magical thinking that adults often exhibit when they try to teach children.

On more than one occasion Piaget has denied the widespread belief that he is a child psychologist. His preference for the description of himself as "genetic epistemologist" is a statement of both his program and his results. He has devoted his life to investigating the epistemological question: How do we get knowledge? Piaget's strategic decision to pursue that question primarily by studying the origins of knowledge in children reflects his belief that knowledge arises through a growth process, that it is constructed by the person in the course of his own adaptive activity. Thus, his focus on the child's constructive cognitive growth in achieving such basic concepts as the permanent object, space, time, and causality is based on a rejection of the *preformist* view that such ideas are innate, acquired by the individual without struggle, simply passed on to him as a cognitive heritage from his progenitors.

At the same time, very early in the course of his investigations, Piaget recognized that the child's cognitive growth does not occur as a passive reflection or copy of external events impinging on his sensory systems, but through his own activity, constantly structuring and restructuring his own schemata, i.e., actively constructing the world he cognizes. Thus, when Piaget speaks of the earliest phase of childhood mentality as the "sensori-motor period" he is not simply describing the brute fact that the child is born with both sensory and motor equipment. Piaget's use of the term "sensori-motor" reflects his rejection of the *empiristic* or copy theory of knowledge and captures his belief that knowledge arises in the organism's activity, not in the passive experience of a sessile sensorium.

This thoroughgoing interpenetration of observation and activity characterizes not only the development of childhood mentality but also another field of human endeavor in which interesting forms of cognitive growth occur, the process of science. Piaget has been sensitive to this aspect of science and has devoted considerable effort to consideration of the history of science as another major domain in which the approach of genetic epistemology might illuminate the growth of knowledge (Piaget, 1967; 1970a; 1970b). This fundamental fact that cognitive processes, or *intelligence*, as Piaget prefers, are growth processes makes it imperative that we devote careful attention to studying their rate of growth and deplorable that psychologists have given so little attention to this question.

Rate of Cognitive Change

I will examine first the rate at which cognitive change goes on in one creative scientist to suggest that the apparent slowness of cognitive change in children is really not unique in children after all. Then I will consider some of the general reasons for this slowness, and the social context in which creative thought goes forward. With this background, I will turn to the details of the development of Darwin's thought, first on a fairly broad sweep covering a period of a decade or so, and then focus attention on a more restricted period of about a year (Gruber and Barrett, 1973).

It is important to underline the question, "What is the rate at which the process of cognitive growth goes on?" Is there any general value in questions about the rate of natural processes? Psychologists often talk as though psychological processes occur out of time, or they are extremely vague about the rate of such processes, treating in the same way events that take place in a matter of minutes and other events that require weeks or even years. This is not the case in all the natural sciences. Turning points in the history of science have often depended on measuring exactly the rate of some fundamental process, such as the velocity of light or the speed of the nerve impulse. We are only a little over a century away from the time at which it was thought that the speed of the nerve impulse, or the speed of thought, if one prefers, was instantaneous and mental processes were therefore miraculous. When Helmholtz in the 1840s measured the speed of neural transmission, he made an important step in a very considered and conscious campaign to put the study of psychological and physiological processes on a material, rather than an idealist, basis (Gruber and Gruber, 1956). The rate of geological change was similarly associated with questions of materialist versus idealist philosophy, as was the rate of evolutionary change. In our times these matters have been more or less settled, at least in very broad outline. Today we are more concerned about the rate of social change, and we have no good theoretical conception of what it is or ought to be. But, with regard to all the other topics that have been mentioned, we can give fairly specific answers. In cases where there is some doubt, such as the rate of evolutionary change, we expect further advances of science that will actually answer the question.

Thus it would be unusual in the pantheon of science if psychology were really to be a field of inquiry that avoided questions of rate or trivialized them in the belief that it really does not matter at what speed fundamental processes go on because it is only the developmental or causal sequence that counts.

We are always interested in rate in any of the sciences not simply because we want to be able to say that dx/dt equals such and such, to record a numerical value, but because we understand that there is some deep connection between the nature of the process and the rate at which it takes place. This is especially important in biological sciences, including psychology. For example, some connection exists between the rate at which one process goes forward and the relation of that process to other processes. In all living systems a number of processes have to be coordinated with each other; if they go forward at one ensemble of rates you will have one outcome, and in another pattern of rates you will get a different outcome. Growth rate is a fundamental question for biological sciences.

Piaget is clearly the foremost individual in making at least some statements about the rate of growth of cognitive structures in children. He has given us not only information about the rate of growth, but also some suggestion as to how it happens that cognitive change should be, roughly speaking, slow rather than fast.

Over a period of years in working on the Darwin material, I came to the conclusion that creative thought, even in Darwin's case, moved fairly slowly. Darwin was a limiting case, a very intelligent man well trained for what he was trying to do. Just as in children's development, we are dealing with the work of construction: an ensemble of processes in time that go on during continuous interaction of the person and his environment.

With respect to adults, large changes in thought and in cognitive orientation usually take considerable time, often, years. Religious and political conversions, or decisions to migrate (as on the part of the Jews who were caught in Nazi Germany)—the total process of groping one's way toward a point of view that permits one to act in a way substantially different from the way in which one might otherwise have acted—seem to take from two to five years. I have found evidence for this in a series of interviews with young Americans who decided to resist military service in the war against Viet Nam. All these instances seem to involve a process of groping, experimenting, discarding one position after another, restructuring ideas, interacting with other people, and so on. The time-scale for change in these adults may not be remarkably different from what has been described for the growth of children's thought or remarkably different from our observations about Darwin. We are probably talking about something fairly general, although the evidence in this chapter comes from the study of one case.

The Process of Change

Why should it be generally true that fairly large changes in the structure of ideas take so much time? Basically, because change is a many-sided process. Every ensemble of ideas is involved with a set of social commitments that the person has made. This is obvious in the examples already cited, and it was true in Darwin's case. The change in thinking means moving away from an established and perhaps hard-won set of relations with other human beings. This may be more important in the case of children than has been realized. When the child, for example, shifts his way of thought so that he restrains himself from making a judgment based on purely perceptual criteria, he is also making a serious change in life-style. He is increasing his independence from the stimulus. In that sense, he is increasing his independence more generally, and any increase in independence carries with it both a promise and a threat. We would probably discover, if we looked a little more closely at those moments when the child's thinking really seems to move, that the child experiences a sense of exhilaration. When we speak of "insight" or the "Aha Experience," it is not just seeing something new. It is feeling. And what the person is feeling is both the promise and the threat of this unknown that is just opening up. When we think new thoughts we really are changing our relations with the world around us, including our social moorings.

A second major point certainly applies in scientific thought. One has to develop a new way of using language—a new vocabulary, new significances for old words, new syntaxes, new ways of putting ideas together. Words that formerly stood for one group of ideas now stand for slightly different groupings. The process of learning a language, or of reconstructing one, takes time.

A third feature of large changes in thought is that they entail changes in the whole structure of an argument, structure meaning a set of ideas that the thinker puts together in some relationship with each other. These relations can change in many ways: Premises become conclusions and conclusions premises; distinctions that seem dichotomous can become graded on a serial scale, etc. As the structure of an argument changes, the very same sentence spoken at two different points in time has a really different meaning. The thinker does not necessarily intend to change the meaning of the sentence, but the context has shifted. As the total argument shifts, the evidence appropriate to support any point in it may shift—the acceptable sources of evidence change. What is a legitimate source, who is to be trusted, what sort of equipment is to be relied upon, what books will be read? The method of mapping fact into theory changes. And the standard of evidence shifts. When there is a brand new idea about which one is very excited, any shred of evidence is welcome and as a result looks very significant. Later on, in the development of thinking, the standards of evidence rise again.

A large change in thought really involves abandoning a paradigm—in other words, abandoning a whole way of thought: a group of ideas, methods, sources of evidence, relationships with colleagues, and so on. Since this change cannot take place instantaneously, every change made in an individual's way of thought moves him away from his own past.

At all times, the changes one is both undergoing and producing are opposed by ongoing vital systems, systems that are personal and idiosyncratic; one's way of perceiving, one's concept of law, one's understanding of the received knowledge of the day. This conflict between the old person and the new one emerging may limit the rate at which one can move.

Another important factor can limit the rate of change. In all the really interesting, major innovations in thought, the person must discipline himself severely by limiting the domain that he is going to attack. He has to expunge certain problems from consideration. An obvious case is that of Isaac Newton. He had no explanation at all for action-at-a-distance, but he did very well by assuming it. He included with his three laws of motion the inverse square law of gravitational attraction. But he had no explanation for gravity, and this troubled him. He spent a good deal of the rest of his life, after those golden years of his twenties, trying to penetrate more deeply into the nature of things so that he could understand gravitational action-at-a-distance, or do away with it, or explain it in terms of the inner structure of matter. Newton having tried and failed, it was essential for him to carve out a domain in which the unexplained premise, gravity, was not in question. He accepted it and went on from that point.

We discipline children, or they discipline themselves, to take a great deal on faith in the course of accepting what we teach them, or thinking for themselves. There is a process in creative thought very like the psychoanalytic notion of the return of the repressed. When the thinking person nears the frontiers of his knowledge, he necessarily touches on a number of primitive, unexplained ideas, some of which he must actually utilize in their unexplained form as premises in his system. The very fact that he is using them is part of the dilemma of the creative person. As a result, he vacillates and backslides. He says, in effect, "Well, I won't try to explain this, I'll just use it . . ." Then he turns around and does try to justify it, at least to himself, after which he returns to his initial stance and moves ahead without justifying every move.

Is all this waste motion? Is it necessary to the maintenance of the cognitive economy? What function does this expenditure of effort play? Even without answering these questions, the creative inability to maintain a posture of unwavering faith in unexplained assumptions clearly helps account for the time one takes to restructure thought.

Social Context of Darwin's Thinking

Darwin's own developmental changes illustrate a few of these points. First, let us consider the social context in which thought takes place. Darwin was dealing with a dangerous idea. Usually, when we think of the suppression of ideas, or the persecution of people for their ideas, we think of it as something that occurs after the fact: You must *have* an idea before you can be persecuted for having it. But if we insist on the view that ideas grow over a period of time and that the person thinking is living in a world and intereacting with it, then he has the opportunity and the occasion and the information necessary to be oppressed during the growth of the idea. That was very much the case with Darwin.[1] Somewhere in his early notes on evolution, about a year before he thought of natural selection, he wrote in a worried tone, "Mention persecution of early astronomers."[2] He went on to say something about the role of the individual in history. He was telling himself to remind his readers about the persecution of Renaissance astronomers by the Church, much of which he had learned from his teachers. Several of them had written on the history of science and were reflective about the process of getting new knowledge. In the course of their reflections, they talked about the oppression of men like Copernicus, Bruno, and Galileo.

Imagine what it would be like to be Galileo, for example, trying to think through a new cosmology, thinking in a world where he would eventually be forced to swear as follows (before the Inquisition):

> I, Galileo, being in my seventieth year, being a prisoner and on my knees, and before your Eminences, having before my eyes the Holy Gospel, which I touch with my hands, abjure, curse, and detest the error and the heresy of the movement of the earth (White, 1955, p. 142).

Such history may seem unreal to us, but it was very real to Darwin for reasons that are quite convincing.

About a hundred years after Galileo's recantation, the French biologist and geologist Buffon had to recant before the Faculty of Theology at the Sorbonne and publish the following:

> I declare that I had no intention to contradict the text of Scripture; that I believe most firmly all therein related about the crea-

tion, both as to order of time and matter of fact; and *I abandon every thing in my book respecting the formation of the earth*, and, generally, all which may be contrary to the narration of Moses (Lyell, 1835, Vol. 1, p. 69).

Bruno had been burned at the stake before Galileo's recantation. Joseph Priestley, the chemist, who was involved with the Darwin family a generation or two before Charles Darwin was born, was forced out of Birmingham in 1791. His house was burned down by a government-inspired mob for his beliefs about questions relating to the French Revolution, religion, and the relation of mind and matter. These events were in the books Darwin read, and they were a part of his family tradition. Perhaps more pointedly, he had had direct personal experience, for he had observed at least one case of the suppression of ideas. When he was a student at Edinburgh a classmate, a Mr. Browne, had read a paper before the student scientific society on the material basis of mind. He had presented a materialist philosophy of mind and life, and it had created such a disturbance that the society had stricken the record of that occasion from its minutes. Fortunately, the secretary of the society made a careful outline of Browne's paper, took a fine pen and ruled one straight line through every line he had written, leaving it very legible.[3]

Yes, Darwin had "been there" in the domain of dangerous ideas. His own grandfather, who was an early evolutionist, had been savagely attacked and lampooned by a quasi-governmental publication for his ideas, and his reputation had suffered severely. The oppression of thought was no empty abstraction to Darwin. He had seen it happen, he knew about it in his own family, and he worried about it.

One of Darwin's dreams seems to reflect his anxiety. In 1837 Darwin began his notebooks on evolution; about a year later he began a second set of notebooks on man, mind, and materialism. In September of 1838, a few days before his reading of Malthus' *Essay on Population*, which helped him to achieve a deep insight into the idea of natural selection, he recorded a dream:

Sept. 21st. Was witty in a dream in a confused manner. Thought that a person was hung and came to life, and then made many jokes about not having run away and having faced death like a hero, and then I had some confused idea of showing scar behind (instead of front) (having changed hanging into his head cut off) as kind of wit showing he had honourable wounds.[4]

The dream seems to show his anxiety about being punished for his ideas (wit), and his hope for immortality (coming back to life); it also shows some concern about the problem of courage ("jokes about not having run away"). Of course, other interpretations of the dream are possible, and there is no reason to think it must be interpreted on only one level. But Darwin's self-instruction in March of 1838 to "mention persecution" and his dream of execution in September are not isolated examples. Another entry in his notebooks says: "To avoid stating how far I believe in materialism, say only that emotions, instincts, degrees of talent, which are hereditary are so because brain of child resembles parent stock."[5] Darwin was always looking for ways of saying things that would soften the opposition.

Out of his concern for the opposition, and the ferocity he rightly knew he could expect of it, came the strategy of deferral and delay that are evident in the chronology of his life. Sometime in 1838, certainly no later than November, when he summarized his views as three principles stated unambiguously, he had a tolerably clear picture of evolution through natural selection. Of course, there were many weaknesses, many loopholes. By 1844, however, he had written the second of his two early essays on evolution; this one he thought was good enough to publish, if he should die prematurely. He wrote instructions to his wife as to how to arrange for editing and publishing them (F. Darwin, 1909). Then he embarked on a long detour, which took 15 years, and finally published the

Origin of Species in 1859. In the *Origin of Species* he said very little, almost nothing about man as part of the evolving web of life, although it is very clear from his manuscripts that he believed throughout this whole period that man had evolved according to the same principles as other animals. It took him another 12 years, until 1871, to write *The Descent of Man*, in which he finally revealed his views on man.

One question that has concerned me, to which only a speculative answer can be given, is whether or not this long detour and deferral in the public expression of his ideas had any effect on the quality of his private thought. My feeling is that the main answer is "no"; on the whole, Darwin was quite honest with himself. There is a distinction between honesty with one's self and openness with the world at large. On the other hand, he might well have been able to move more rapidly and smoothly, with less backsliding, if he had not been so obsessed with justifying every step that he took. If he had not been frightened, he might not have been so obsessed. In fact, of course, he was obsessed, he was frightened, and he did compulsively collect tremendous amounts of material.

There is a second way in which he must have suffered from secrecy. Piagetians recognize the importance of dialogue, of children's discussing and arguing with each other as playing a role in the development of their thought. Darwin did have a few confidants with whom he could share his thinking on certain subdivisions of his great enterprise, but few with whom he discussed the whole picture. The one person with whom he could be most open, the botanist Joseph Hooker, was often away in Asia. Certainly, the necessity for secrecy, as Darwin felt it, prevented him from opening up to other people in ways that might have been helpful for the development of his ideas.

The suppression of dangerous ideas may seem far afield in a discussion of children's thinking. But it is not. Children grow up in a world full of taboos. Has any child in a hygiene class ever asked his teacher, "How do you wipe yourself so you keep your hands clean?" Or consider the child's ideas about digestion. This subject provides the child with a marvelous opportunity to think about very interesting processes involving the conservation of matter during various mechanical and geographical changes. Children do think about this subject, because it concerns their own bodies and certain activities of great interest to them. Can anyone imagine that the child's thinking is unaffected by the taboos surrounding discussion of the movement of food from the slaughterhouse to the toilet?

A few years ago Danielle Naville, William Walsh, and I studied children's ideas about digestion. We investigated the question purely from the point of view of cognitive development. We were aware of the taboo aspects of the subject but attended to them only to the extent of carefully establishing good rapport with the children. With the present considerations in mind it becomes important to examine more closely the possible effects of selective repugnance in leading to uneven development in the child's thought. Think of the enormous number of restrictions on thought, and their great power in cases involving taboos concerning not only the body, but types of social relationships and ideas about God, politics, and authority. A cognitive psychologist who ignores such questions is himself a case in point, a scientist keeping himself ignorant of the facts of life.

Another feature of the social context in Darwin's development was a set of relationships he had with his university teachers, first at Edinburgh and then at Cambridge. He somehow sidestepped the oppressive effects of an autocratic lecture system, and benefited from a very personal kind of encounter with a few teachers. From the age of 17 on, he had a strong and warm relationship with these professors, went on field trips with them, and worked in their laboratories. He was a junior collaborator of several distinguished people. The most obvious result was good training for independent scientific work. Another result, a little harder to see, was that he really cared about his teachers as individuals; he cared about how they would feel if he expressed thoughts offensive to them. These Cambridge professors were, by and large, transitional figures in the history of science. They worked at a moment when it seemed important to do the best one could to bring religion and science to terms with each other without conflict. It was likewise important to them to give up outmoded ways of thought at a dignified rate, fast enough to assimilate the undeniable advances of science, slow enough to avoid unsettling the establishment. They counseled moderation.

Charles Lyell, who became one of Darwin's most important mentors, understood the value of moderation as a strategic device: If one offends the people one is trying to persuade, one won't be able to persuade them. In writing his classic *Principles of Geology*, Lyell certainly thought of the earth as millions of years old, perhaps infinitely old. But in his book, when he wanted to convey the idea "very old," he used one phrase repeatedly, "twenty thousand years or more." In a geological sense, twenty thousand years may seem ridiculously short to us. But it was more than three times as old as the Biblical creation, as it was then understood. Lyell carefully chose a length of time that would jar his contemporaries but one that they could comprehend. He wrote, in a letter to a friend,

> If we don't irritate . . . we shall carry all with us. If you don't triumph over them, but compliment the liberality and candor of the age, the bishops and enlightened saints will join us in despising the ancient and modern physico-theologians . . . If I have said more than some will like, I give you my word that full *half* of my history and comments was cut out, and even many facts; because either I, or Stokes, or Broderip, felt that it was anticipating twenty years of the march of honest feelings to declare it undisguisedly. Nor did I dare to come down to modern offenders (Lyell, 1881).

The last sentence in Lyell's sketch of the history of geology indicates that he stopped with Buffon, who had worked about

90 years before him. By keeping his history disconnected from the present he depersonalized the suppression of ideas, making it easier for his contemporaries to assimilate his historical sketch without rejecting it because of its painful personal relevance. Lyell is the man who introduced Darwin to the London scene. He taught him the scientific diplomacy that led to his strategy of delay.

When we turn to the contents of Darwin's thought about evolution, we ought to keep in mind this risky side of creative thought. Darwin was not working in a vacuum where ideas and facts were all that counted. He sensed the significance of his ideas for the age, and the effect of saying them in a certain way on the people he loved. He really did love them. The emotional attachment was part of the thought process, a part of the growth of thought. The loyalties and the fears, the sense of tension between the old and the new, the fear of enemy ridicule and the even stronger reprisal of withdrawal of a friend's support—these are far more pervasive forces in the history of ideas than is recognized in our cognitively oriented discussions. Perhaps we can afford to flatter ourselves that there are no more dangerous ideas, that nowadays thought is free. I question this.

I shall now look more deeply into the contents of Darwin's thought and examine the rate of change of his ideas, beginning with the period of the *Beagle* voyage, 1831–1836. Much of the time Darwin was not on board ship; it docked and he went on horseback or on foot all over South America. The ship visited islands, first in the Atlantic and later in the Pacific, and he was the naturalist of the ship. His job was to explore and observe widely, to collect zoological, botanical, and geological specimens, and to interpret his findings. (Others on board were concerned with more strictly naval matters, especially measuring the longitude.) During that period, Darwin went through a number of distinct phases. When he began the voyage, he was persuaded along with most of his contemporaries that there had been a Creator who had ordained a physical world and a set of organisms to inhabit it. In Figure 4–1 causal arrows are shown from the Creator to the world and to the

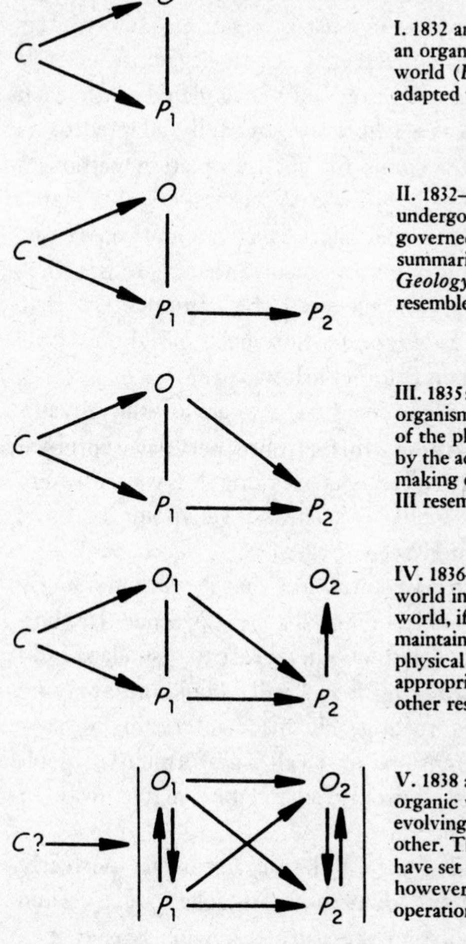

I. 1832 and before: The Creator made an organic world (*O*) and a physical world (*P*): *O* was perfectly adapted to *P*.

II. 1832–1834: The physical world undergoes continuous change, governed by natural laws as summarized in Lyell's *Principles of Geology*. In other respects, II resembles I.

III. 1835: The activities of living organisms contribute to the evolution of the physical world, as exemplified by the action of the coral organism in making coral reefs. In other respects, III resembles II.

IV. 1836–1837: Changes in the physical world imply changes in the organic world, if adaptation is to be maintained; the direct action of the physical milieu induces the appropriate biological adaptations. In other respects, IV resembles III.

V. 1838 and after: The physical and organic worlds are both continuously evolving and interacting with each other. The Creator, if one exists, may have set the natural system into being; however, He does not interfere with its operation but stands *outside* the system.

FIGURE 4–1
Darwin's Changing World View

organisms, to indicate that they were His creations. The organisms were beautifully adapted to their environment, not through any direct causal relationship between organism and environment, but because He created them so.

An important idea in Darwin's day was the *argument from design* for the existence of God; that is to say, if one considers an organism that works as perfectly as any organism we may study, all of its organs so beautifully articulated with each other and the organism as a whole so beautifully adapted to its world, how might one account for such a creation without a Creator, a divine Artificer, a divine Watchmaker? If a watch is found, one must believe that there is a watchmaker. If one finds a perfect organism, one must believe that there is a Creator. Darwin seems to have accepted this argument in 1830 and described in his *Autobiography* how he admired one of its chief proponents, William Paley (Barlow, 1958).

In the period 1832–1834, the first 2 years of the voyage, Darwin's position shifted; he came to recognize that the physical world was not static. He was working his way through Lyellian geology—the modern, scientific viewpoint that the physical universe had not been created some 6,000 years ago and then remained relatively static, but that it was indefinitely old and continuously undergoing change governed by uniform natural laws. It took him about 2 years to assimilate fully that position. He was studying it in Lyell's book and applying it to the places he was visiting. He made interesting discoveries and important extensions of Lyell's uniformitarian geology, but basically he was simply assimilating a new way of thought, a new paradigm.[6]

Let us anticipate a little bit. If the organisms are perfectly adapted to their world and then the world changes, a certain disequilibrium is implied. Eventually Darwin saw that a change in the world implies a change in organisms, if adaptation is to be maintained. The underlying premise remains adaptation, which was also the underlying premise in 1830. But Darwin did not move directly from a belief in a static physical world with a static set of well-adapted organisms to his belief in a world in flux. Instead, he became preoccupied with working out a particular application of uniformitarian geology to the theory of the formation of coral reefs. The basic point is that a small organism is changing the physical character of the world. The skeletons of the coral organism, as they grow and die and accumulate, form a coral reef. Darwin worked out the exact way in which this happens. It was one of his really elegant scientific successes at that time. It was also important in a personal way, in helping him to grow and to see that, although Lyell was his model and his mentor, he could surpass him. Darwin's theory was at variance with Lyell's theory of coral reefs, but Lyell immediately accepted Darwin's theory and congratulated him on it (Gruber and Gruber, 1962).

In looking at the broad architecture of Darwin's thought, his coral theory says that the organisms existing at one moment in time produce changes in their physical environment. As one gets more and more complex relationships among changing organisms and their changing environments, the Creator stands further and further outside the system of nature. One can put all the arrows into the diagram if one likes, and one can find some group of facts exemplifying each one of these relationships. They are all mutual and bidirectional and form a system of living nature. One can say that the Creator set the whole thing in motion. In later life, for public consumption, Darwin still conceded the possibility of an Originating Creator, but not an Interfering One. The former would stand outside the system of nature and would not compete with natural processes; the latter would amount to a rival hypothesis, vying with natural selection in its ability to explain specific biological phenomena. The disparity between

Darwin's private agnosticism and his publicly expressed ideas have long confused certain issues in his life, which new manuscript discoveries are now making clear.

Let us consider the position he had early in 1837. He said in his notebook, in effect: If the world changes, organisms must. At this point he began to think seriously about evolution. One may speculate that he was thinking about evolution during the entire voyage of the *Beagle.* The fact remains that in 2,000 pages of geological and zoological notes that Darwin kept during the voyage there is very little about the evolution of organisms. What little there is actually denies the possibility. Darwin was cognizant of previous theories of evolution, but he did not discuss the subject at all in the *Beagle* notes. One might say that he thought about it but did not happen to put it down. But there is no evidence of this, and there is much positive evidence that when Darwin thought about a subject his ideas appeared promptly in his inexorable notebooks.

Psychologists who have not been imbued with the idea of slow cognitive growth may wish to believe that Darwin had one sudden insight into the whole theory of evolution through natural selection and merely wrote it down later. On this point we have clear evidence that forces us to abandon any such notion of sudden insight in favor of a growth model. In July of 1837, when he began his notebooks on evolution, his point of view was very different from that in his eventual theory. Indeed he had a theory, and he had what felt to him like a great insight. His sentences often ended with exclamation points, and he wrote with an air of excitement. A few pages later he got another insight. His notebooks were full of them. But the notion that his first effort was *the* insight is misleading, as would be any suggestion that these early efforts are partial insights, all to be accumulated as part of the eventual theory. Rather, each move is a step along the way in a complex growth process. Some early ideas can be discerned as elements in the eventual product, and some must be discarded.

This kind of growth is a movement from structure to structure, and the fate of any element as well as its meaning depends on the structure of which it is a part.

The theory of evolution that Darwin began with in July of 1837 stemmed in good part from Leibnitz and Lamarck: The world is permeated with monads, small particles that can spring into life through some undetermined cause. Once alive they begin to evolve. This process of monads coming into existence or beginning to evolve has been going on all the time. Unlike Lamarck, Darwin had a conception that there was not an infinite number of species. There were some limits to the variety to be found in nature. Consequently, if monads are growing and evolving all the time, there must be some way of killing them off in order to limit the variety extant in nature, to preserve the approximate number of species. Lamarck had believed that there was no such thing; if some creature seemed to be extinct, he argued, all you had to do was look harder and you would find it somewhere, perhaps in the unexplored realms of the world.

Darwin, however, needed a mechanism of extinction. His notion in July of 1837 was that the monad quite naturally should have a limited lifespan. Individual organisms have lifespans, so why not monads? When the monad dies, then all the species it has become in its evolution must also die. This was the mechanism of extinction that he posited in July of 1837. It is a strange notion to us, and it had little in common with Darwin's later thought. Interestingly, a number of scholars have read Darwin's transmutation notebooks without noticing the monad theory, although Darwin plainly spelled it out. If one consults the only published index of Darwin's notebooks, the word *monad* does not appear. It does appear in the text. And the same individual who prepared the index was involved in transcribing the notes and approved the use of the word *monad* plus a whole set of synonyms Darwin used for it (de Beer, 1960–1967). But an idea that has become outmoded is nearly invisible unless there is an epigenetic view of the growth of ideas.[7]

It is difficult to believe that Darwin, the hard-nosed, empiricist scientist whom we think of as the Darwin of the *Origin of Species*, once believed in monads, continuing spontaneous generation, and species lifespan. Nevertheless, in July of 1837 he had a set of ideas that had its own evolution as I have sketched it. It was richly determined by the great amount of material he had at his disposal and by his complex background; it was a network of ideas involving adaptation, the functions of the reproductive system, taxonomy, and extinction.[8] It was not a mere collection, but a network of interrelated ideas, in a set of compensatory and mutually supportive relationships. Although some of this initial theory was "wrong" in the sense that he had to abandon it to develop his later theory, all of it may have been "right" in that it helped him to move in that direction.

It is hard in studying any cognitive growth process to be sure how many stages there are. But we can certainly discern in Darwin's thinking in the 15 months from July 1837 to September 1838 other interesting models describable with about the same coherence as the monad model, although I will not describe them here.

In September of 1838 Darwin experienced his famous insight on reading Thomas Malthus' *Essay on Population*. The notebook passage is full of inserts and written with excitement. It reads in part, "Population is increase at geometrical ratio in FAR SHORTER time than 25 years—yet until the one sentence of Malthus no one clearly perceived the great check amongst men."[9] The basic point is that nature has the capacity of producing many more organisms than could possibly themselves survive and reproduce. If each one did, imagine how many children one woman could have, not to speak of other organisms that reproduce much more rapidly than *homo sapiens*. Benjamin Franklin had calculated that Englishmen would soon outweigh the earth if their reproduction went unchecked; Malthus cited Franklin; Darwin patriotically changed the instructive example from Englishmen to elephants. In the notebook passage of September 25, 1838, Darwin emphasized food supply as being the major check. Later on he expanded that idea to include all the different ways in which population growth is limited.

In September of 1838 he seemed finally to have clearly formulated the idea of natural selection as an evolutionary mechanism. This was one week after his dream of decapitation. He did not immediately say, "This is it. I've got the great idea that I've been looking for." Instead, he reverted to a whole set of other complex concerns that have to do with his search for the mechanism of variation, a preoccupation of his that merits careful attention.

About 3 months later, in December 1838, he recapitulated his theory of evolution through natural selection. He wrote:

Three principles will account for all
(1) Grandchildren like grandfathers
(2) Tendency to small change especially with physical change
(3) Great fertility in proportion to support of parents.[10]

The first point is the common observation that like breeds like; the last point is the Malthusian superfecundity principle. The second point is the most interesting. Although he says "Three principles will account for all," he actually has at least four principles in his list, because he would not discipline himself to assert the premise, "tendency to variation," without importing into it something about the *cause* of variation. Because the unexplained premise troubled him, he supplemented it with the thought, "especially with physical change"—that is, the tendency to vary might be at least partially accounted for by variations in each organism's immediate physical environment.

To derive natural selection as a necessary consequence those three points are essential: like breeding like, frequent small variations, and the Malthusian principle of superfecundity or population pressure. As Darwin later recognized, additional theoretical steps are necessary in order to advance from these three principles to divergent, progressive evolution. But for the moment Darwin thought they would "account for all." Except that there were really four: The phrase ". . . especially with physical change" is an example of the backsliding referred to earlier. He really knew from about this point on that to assume variation without explaining it would be a workable strategy. But it is incredibly difficult to make a cardinal principle of something and then deliberately not to explain it. It takes a lot of courage to expose one's ideas without having an explanation for the primitive underlying premises. In the *Origin of Species*, written 21 years after his notebook entry, Darwin wavered in the same way. The opening chapter contained a feeble attempt to explain the cause of variation, in which he concluded, "The result of the various, quite unknown, or dimly seen laws of variation is infinitely complex and diversified" (Darwin, 1859, p. 12). Although it begins with this unexplained premise, the rest of the book is a forceful work, a great classic of science writing. Logically, he might well have begun simply by asserting and documenting the *fact* of widespread variation without trying to explain it. The approach he took shows how stressful it is to expunge a problem in order to think constructively about those things that are attainable for you.

Throughout Darwin's notes and in some of his published writings as well one finds him grappling with the causes of variation and then throwing up his hands and saying in effect, "Well, I can't really handle it. But I can do this . . .," after which he does whatever he does successfully. This tendency in Darwin has sometimes been seized upon to depict him as a weak and vacillating man who wavered between Lamarckism and Darwinism, falling back on the former whenever the principle of natural selection was criticized. On the contrary, insofar as the handling of the whole issue reflects on Darwin's personal character, it shows one of his strengths. He was able to recognize that he was using an unexplained premise, to preserve his creative discomfort with that difficulty, and still to forge ahead to what could be done. Insofar as the issue reflects upon the social character of science, it adds weight to the earlier observations about the operation of oppressive forces during the ongoing process of creative work. If there had been a more welcoming attitude toward novelty, if Darwin had been able to share his dilemmas with others, it may well be that someone might have seen sooner and more clearly how the argument could be divided into two parts, the problem of evolution and the problem of variation.

A large part of the process of the growth of ideas has to do with the changing structure of an argument, so that an idea having one meaning in one context has quite a different meaning or a different thrust in another context. This point applies both to understanding the growth of one person's thought and to comparing individuals with each other. For Lamarck, a cardinal principle was the following causal sequence: Changing circumstances lead to change in activity, lead to change in structure. It was the only *causal* theory of evolution Lamarck had, since he lacked the idea of natural selection. He had another, *non*causal principle of evolution—that there is in nature an inherent tendency toward progressive change. Darwin considered the latter idea too metaphysical and rejected it.

For Darwin the same causal sequence (changing circumstances lead to change in activity, lead to change in structure), as a hypothesis about the causes of variation, was a kind of grace note on a strong theory that could survive quite well without it. In fact it still does. Even today we do not know the causes of mutation. That, however, does not make us less Darwinian in our thinking.

Viewing Darwin's thought as a dynamic structure of ideas,

we can better understand that it was difficult for him to move forward on the main line of his evolutionary thought until he could reduce the amount of effort he put into the search for the mechanism of variation. A number of possible mechanisms interested him: hybridization, geographical isolation, and simply environmental influences. The last he thought of as operating in two ways: first of all, directly, somehow or other, on the organism; and second, by influencing the activity of the organism in ways that would lead to changes in its structure. The search for the causes of variation was evidently a complex and fascinating distraction, one not easy to escape.

There is another fascinating aspect of Darwin's early reflections on the relationships among environment, activity, and structural change. Darwin began his notebooks on man, mind, and materialism in July 1838. In other words, 2 months before he clearly formulated the principle of evolution through natural selection, he went off on what seems to be another line of attack. He did not operate in a linear fashion, first solving the great problem of the cause of evolution and then applying his solution to man. He began a separate set of notes and a systematic attack on the problem of man and mind while only half-way along in his search for a general theory of evolution.

There are probably two main reasons for his beginning the notebooks on man, mind, and materialism in parallel with the notebooks on evolution. First, he needed to explore the appropriate scope for an evolutionary theory. Could he treat man as part of the natural process of evolution, and man's brain as the organ of mind, evolving according to the same laws as any other living matter? Second, he needed to find the cause of variation, an enterprise for which the human brain, so modifiable an organ, seemed to provide promising material. Darwin was attracted to the idea that when the environment exacts different activity from the organism, the change in activity leads to structural change; that is, a habit could become a structural change, or the brain was the material storehouse of changes in mental functioning—the brain was the organ of mind.

Darwin had become a materialist. He believed that habits, dreams, and old memories that can be suddenly revived all have a material existence in the storehouse of the brain. The survival of an idea was for him both evidence of a structural change in the brain and also the cause of inheritable structural change. Human learning and experience could be viewed as structural modification through behavior and thought, and they were consequently suitable ground for studying the causes of variation; the mind is the most modifiable function and therefore the brain must be the most modifiable organ of the body.

The hypothesis that activity produces heritable modification has different meanings in different contexts. Within the framework of biological thought it is simply an idea about the cause of variation. But in a larger philosophical and theological context, the bold declaration that the brain is the organ of thought and that it evolves like all other organs has import for any discussion of man's place in nature, of his relation to any supposed supernatural agency, and of the whole issue of philosophical materialism. Darwin was sensitive to these ramifications. He knew that he was thinking dangerous thoughts.

The vicissitudes of thought on the time-scale of history help to emphasize the point that one really can make choices while he thinks. A problem in its particular shape has "demand character" that draws the thinker on—structural requirements that must be met, as Gestalt psychologists have made clear. But there is always more than one possible path.

Divergent paths are taken even over a short span of years. For example, a striking historical change has occurred in the interpretation of evidence about experiments in cognitive development. At some point, perhaps 10 years ago, the failure of training experiments to produce changes in children's cognitive development was taken as evidence for a Piagetian point of view and against an empiricist, copy-theory point of view. Piagetian training experiments detailed in Chapter 3 showed that one cannot simply shove facts in front of children and expect them to profit from the experience. The child must have the requisite level of cognitive development in order to take the next step called for by the teaching adult. Today evidence for successful training experiments is taken as support for a Piagetian interactionist position. The contradiction between these two historical phases is more apparent than real. There have been theoretical developments and corresponding changes in method, so that both the earlier and later interpretations can be correct and noncontradictory. But looking through a sufficiently coarse lens, we might see a contradiction.

Something similar happened in the history of evolutionary thought, in connection with the idea of adaptation. In the 1820s, when Darwin was reading Paley's works, adaptation was taken as one of the leading sources of evidence for the existence of a Creator and a static universe in which all things were perfectly adapted to each other. The Reverend William Paley, the great proponent of the *argument from design*, gloried in the beautiful adaptations to be found throughout living nature, and likewise did Darwin's teachers. Natural Theology meant the study of nature as a way of glorifying God's creation. Fifty years later, after Darwin's theory had been victorious, a wave of enthusiasm erupted for discoveries of new and subtle adaptations. Every new adaptation that was discovered and clearly interpreted was taken as evidence of the Darwinian theory of evolution. It would appear that we have to look very carefully at the structure of an argument at each moment in time. Where are we? What is it that we do not know? Where are we trying to go? The facts we deal with at a given moment are not facts in isolation; they are facts that are discovered, assimilated, and presented in relation to a set of ideas, a set of questions, a set of choices facing us.

At every such juncture we may choose the pathway that leads to the minimum disturbance of pre-existing ways of thought, or we may choose to move in other directions.

Such choices were open to Darwin and he often took the harder, riskier paths. Obviously, he might have avoided the whole subject of evolution, for at the time he took it up he was already well launched on a successful career in geology. Having accepted the challenge to develop a theory of evolution, he might well have stopped short of man, or at least avoided discussing the evolution of those features of human existence that seemed to his contemporaries to set us off from other animals—our moral sentiments and our higher intellectual powers. But Darwin chose to consider the whole human being. Nothing was too sacred, nothing would be permitted to escape from the web of evolutionary theory.

We may say, then, that Darwin's interest in the relation between individual function and inherited form had two main sources: his desire to explain the causes of variation and his desire to go as far as possible to include the highest human psychological faculties in his system of evolutionary thought. It is true that Darwin showed much moderation and strategic caution in deferring the public exposition of his ideas. But in his private thinking and ultimately in his published works he had the courage to explore new paths in human thought and to travel them as far as he could go.[11]

To return to the question with which we started: How long did it take Darwin to think out the theory of evolution through natural selection? It might be fair to include in this thought process the 5 years of the *Beagle* voyage, 1831–1836, as an immediate prologue, because in this period he assimilated the conception of a continuously evolving geological system. We shall certainly include the 15 months from July 1837 to September 1838, from the beginning of his evolution notebooks to his insight on reading Malthus, and we might go on a little further, since a thought process has no definite stopping point. This would give us a minimum duration of 15 months and a maximum of about 7 years. The much longer time until the publication of the *Origin* (1859) and *Descent* (1871) does not represent the time to think so much as the time to enact his strategy of voluminous documentation and of delay in the publication of his ideas.

The long delay between conception and publication reflects his reaction to the oppression of thought rather than the time it took to think his theory through. We have seen that even on the shortest plausible time-scale Darwin was constantly troubled by the taboo character of his ideas. Throughout the whole thought process he experienced some anxiety. But in his inner intellectual life he did not yield to these pressures. It is a matter for speculation whether his creative thinking might have gone forward more rapidly under less oppressive conditions. One hopes.

In discussions of the educational significance of Piaget's work, three distinct voices are heard. One stresses the fixity of the order of development and concludes that the adult must be extremely attentive to each child's particular mental level in order to provide him with optimal growth experiences. A second voice stresses the major conceptual areas Piaget has investigated—number, space, time, causality, chance, morality, etc.—and searches for the best teaching materials for facilitating cognitive growth in each area. The third voice stresses the spontaneous, self-guided, active interaction with environment by which the inner mental growth of each child is nurtured and concludes that the teacher's main task is to foster conditions under which each child can think freely.

Piaget's investigation of the stages of children's thinking and his insistence that the order of development is fixed but the rate is variable have had, in some quarters, what may be an unfortunate effect. He has given those who are blindly interested in accelerating the child's growth an estimable target, a set of concepts that may become the "curriculum" of some erstwhile Piagetian school. But in the rush to accelerate growth, one should remember that overexpectancy can be a form of oppression. Piaget would certainly agree that each form of thought must be understood and valued in its own right. Childlike thoughts seem silly or even dangerous because children's thinking is not easy for adults to understand. Childlike thoughts invite laughter, which may be experienced by the child as ridicule. The adult's desire to rush the child into adult ways of thought may be experienced by the child as lack of respect for him as he is.

One day I found my daughter playing with her shadow. She seemed to be trying to get into a room and close the door, leaving the shadow outside. She was having trouble because the light source was a window inside the room. Not wanting to seem silly, she refused to explain her game.

How can we create a world in which a childlike thought will be treated with the respect it deserves? In which the child will know he has that respect? Perhaps this is the right way to read Piaget's work for its educational significance—not as a fixed chronicle of stages in the emergence of a specific inventory of concepts, but as the model of a man who respects children's thinking. Their thinking, like Darwin's, is "childlike"—questioning, searching, following unexpected leads, inventing, discovering. Their thinking is creative, especially when it is permitted to function freely and is respected as all such thought ought to be.

NOTES

1. I have consulted the original Darwin manuscripts in the Cambridge University Library. I wish to thank Mr. Peter J. Gautrey and Dr. Sydney Smith for their help in connection with the use of these manuscripts. Darwin's transmutation notebooks have been published. See "Darwin's Notebooks on Transmutation of Species," edited with an introduction and notes by Sir Gavin de Beer. In the following notes I refer directly to the original manuscript pages. See the reference section of this chapter for the full citation.

2. Darwin, C. Second notebook on transmutation of species (February to July, 1838), p. 123. Darwin Papers, Cambridge University Library.

3. Minute Book of the Plinian Society, manuscript collection, University of Edinburgh Library.

4. In 1838–1839 Darwin kept 2 notebooks on man, mind, and materialism, which he labeled "M" and "N." The dream of execution is recorded on p. 143 of the "M-notebook." These manuscripts are printed in Gruber and Barrett, *Darwin on man: A psychological study of creativity*. New York: E. P. Dutton, 1973.

5. M-notebook, p. 57 (see note 4 of this note section).

6. For a fuller discussion of Darwin's thinking during the *Beagle* voyage see H. E. Gruber and V. Gruber (reference section of this chapter) for the full citation.

7. The author has found something similar in interviewing conscientious objectors. It is difficult to remember what it was like *not* to be a pacifist. Having become one, the person finds it very hard to understand that once he would have been prepared to kill another person.

8. Darwin, C. First notebook on transmutation of species (July 1837 to February 1838), Darwin Papers, Cambridge University Library, pp. 1–35 *passim.*

9. Darwin, C. Third notebook on transmutation of species (July 15, 1838 to October 2, 1838), Darwin Papers, Cambridge University Library, pp. 134–135.

10. Darwin, C. Fourth notebook on transmutation of species (October 1838 to July 1839), Darwin Papers, Cambridge University Library, p. 58.

11. For a fuller discussion of the inter-relations between Darwin's theory of evolution and his materialistic philosophy see Gruber and Barrett, *Darwin on man: A psychological study of creativity*. New York: E. P. Dutton, 1973.

REFERENCES

Barlow, N. (ed.). *Charles Darwin, Autobiography*. London: Collins, 1958.

Darwin, C. *On the origin of species by means of natural selection*. London: John Murray, 1859.

Darwin, C. *The descent of man and selection in relation to sex*. London: John Murray, 1871.

Darwin, F. (ed.). *Charles Darwin, The foundations of the origin of species: Two essays written in 1842 and 1844*. Cambridge, England: Cambridge University Press, 1909.

de Beer, G. (ed.). Darwin's notebooks on transmutation of species, *Bulletin of the British Museum (Natural History), Historical Series*, Vol. 2, No. 2, 1960; Vol. 2, No. 3, 1960; Vol. 2, No. 4, 1960; Vol. 2, No. 5, 1960; Vol. 2, No. 6, 1960 (with M. J. Rowlands); Vol. 3, No. 5, 1967 (with M. J. Rowlands and B. M. Skramovsky).

Gruber, H. E. and Gruber, V. Hermann von Helmholtz: Nineteenth-century polymorph, *Scientific Monthly* 83 (1956), 92–99.

Gruber, H. E. and Gruber, V. The eye of reason: Darwin's development during the *Beagle* voyage, *Isis* 53 (1962), 186–200.

Gruber, H. E. and Barrett, P. H. *Darwin on man: A psychological study of creativity*. New York: E. P. Dutton, 1973.

Lyell, C. *Principles of geology: Being an inquiry how far the former changes of the earth's surface are referable to causes now in operation.* 4th ed. 4 Vols., London: John Murray, 1835. Vol. 1.
Lyell, Mrs. (ed.). *Life, letters and journals.* London: John Murray, 1881.
Piaget, J. *The child's conception of physical causality.* New York: Harcourt Brace Jovanovich, 1930.
Piaget, J. *Logique et Connaissance Scientifique.* Paris: Gallimard, 1967.
Piaget, J. *Genetic epistemology.* New York: Columbia University Press, 1970(a).
Piaget, J. *Structuralism.* New York: Basic Books, 1970(b).
White, A. D. *A history of the warfare of science with theology in Christendom.* New York: George Braziller, 1955 (orig. 1895).

PART II

The Developing Child

What is childhood then? And how are we to adjust our educational techniques to beings at once so like and yet so unlike us? Childhood . . . is not a necessary evil: it is a biologically useful phase whose significance is that of progressive adaptation to a physical and social environment.

(Piaget, 1970, p. 153)

EDITORS' INTRODUCTION

The study of adaptation in the child is one of continuing differentiation of self and constant elaboration of the universe. Forms of intellectual activities are constructed from the reflexes of the newborn on the sensori-motor level. The infant first experiments with objects and with relationships between objects. He advances by about the time he is 2 years old toward objectification of reality by being able to extend, reorganize, and coordinate existing behavioral sequences into new combinations of actions. For example, first he sees a rattle; then sees and touches it (by chance); then sees and touches intentionally and repeatedly (to observe the motion or hear the sound); then sees, grasps, and brings it to his mouth. Initially, he disregards the rattle entirely if it is moved out of his direct field of vision or if it drops to the floor. Later, if he has been manipulating it and it drops to the floor, he may follow it with his eyes, but only momentarily, and then turn to something else. Eventually, he will actively search for the rattle even when it is hidden from his immediate sight, though only in the place he first observed it being concealed. After a time he will search in successive places for it. Around 18 to 24 months he will have achieved the capacity to symbolize his overt actions. He can imitate objects, utilize them in new kinds of play endeavors, and articulate space, time, and causal relationships. According to Piaget he has attained a first landmark in intellectual development—that of object permanence.

Birns and Golden in Chapter 5 review the importance of

Piaget's contribution to our understanding of the infancy period. They describe observations of cognitive development made on children between 12 and 14 months of age from different social classes. The authors make a distinction between certain qualitative differences in level of intellectual functioning before language occurs, when the child's knowledge of his environment is achieved spontaneously through his exploration and experiences, and after the onset of language, around 18 months. At this time, they postulate, the verbal interactions of adults surrounding the child may begin to have an effect on intellectual development. Birns and Golden conclude with certain recommendations for persons working with infants in day-care centers and nurseries.

During the preoperational stage from approximately 2 to 7 years of age, the young child's intellectual functioning is limited to his overt actions. His thinking resembles that of a slow-motion film, according to Piaget (1962), representing one static frame after another but lacking a simultaneous, encompassing view of all the frames. The young child shows tendencies of egocentrism; a pronounced characteristic to focus on one striking feature of an object to the neglect of other important features; and an inclination to see the result of an action rather than the process of the action—all making for inflexibility or rigidities in his thinking. During this stage, social exchange with other children and adults, language, and the maturational development of the child interact to enable the child to contemplate or begin to contemplate his actions, to get beyond the present of his concrete acts, and to start codifying symbols that make possible a sharing of his ideas.

Duckworth in Chapter 6 pays particular attention to the relationship of children's language and linguistic forms to their thought. She traces briefly the evolution of intelligent behaviors achieved by the infant prior to the emergence of representation, that is, before he is able to hold on to an image or idea of an object when it is not immediately present. She then deals with the ways a child begins to acquire representational ability through his play, dreams, imitation, and beginning use of words. Finally, Duckworth looks at the relationship of logical understanding to language, pointing out that children's familiarity and use of certain words often do not include the same grasp or understanding of the words that adults have. She concludes with the suggestion that teachers should be concerned less with the conventional forms of language skills and more with assisting children to develop understanding of meanings. She uses an illustration in the teaching of spelling to amplify this point.

An important concept that must not be overlooked in this description of the developing child is the crucial role played by assimilation and accommodation, which are viewed by Piaget as twin mechanisms of intellectual adaptation. The toddler who plays in the sand, repeatedly filling a pail, emptying it out, and refilling it, is engaged in predominantly assimilative activity, that is, "bending a reality event to the templet of his ongoing structure" (Flavell, 1963, p. 48). When the child later uses the same materials as cooking utensils in sociodramatic play with a companion, the process is preponderantly accommodative in that he is imitating what he has observed —he has reorganized his own earlier actions and converted them into a more complex and differentiated structure. Piaget (1954) views the assimilation process as the more conservative, providing the gradualness and continuity of intellectual development; whereas accommodatory acts result in the child's continually extending and applying new concepts and actions to new and different experiences. The developmental changes that occur in the child's construction of knowledge come about, in one sense, as he searches to regulate these mechanisms of assimilation and accommodation, to consolidate what he already has internalized, and to readjust to new stimuli that disturb existing internalizations. Langer (1969) describes this process of regulation or equilibration as follows:

Our own view is that the child is an active operator whose actions are the prime generator of his psychological development. When he is in a relatively equilibrated state, he will not tend to change; he will only change if he feels, consciously or unconsciously, that something is wrong. This means that both affective and organizational disequilibrium are necessary conditions for development. When these conditions are present, the energetic or emotional force for change in action is activated, and stabilizing interactions between mental actions and the symbolic media in which they are represented can be constructed in order to generate greater equilibrium. It is this constructive activity that constitutes the force of self-development (p. 36).

The absence of constructive activity to which Langer refers can also hamper the teacher. Voyat in Chapter 7 suggests that the imposition of a ready-made curriculum or a predetermined set of goals on the teacher does not allow the teacher to construct her own understandings of what she is attempting to do and how best to do it. Voyat describes in detail the child's acquisition of seriation as an illustration of assimilation and accommodation, or more specifically, as an illustration of the interactions of psychological factors with the internal structure of the child which lead to cognitive transformations. The development of operativity in the child is one of the landmarks in progress the teacher needs to be able to identify and to understand if she is to assist the child at his particular level of development. Voyat concludes by pointing to the fundamental necessity for respect of both the teacher's creativity as well as the child's spontaneity in the learning process.

REFERENCES

Flavell, J. H. *The developmental psychology of Jean Piaget*. New York: Van Nostrand Reinhold, 1963.

Langer, J. Disequilibrium as a source of development. In P. H. Mussen, J. Langer, and M. Covington (eds.), *Trends and issues in developmental psychology*. New York: Holt, Rinehart, and Winston, 1969.

Piaget, J. *The construction of reality in the child.* New York: Basic Books, 1954; London: Routledge and Kegan Paul, 1955.
Piaget, J. *Play, dreams, and imitation in childhood.* New York: Norton, 1962; London: Routledge and Kegan Paul, 1951.
Piaget, J. *Science of education and the psychology of the child.* New York: Orion Press, 1970; Paris: Denoel, 1969.

CHAPTER 5

THE IMPLICATIONS OF PIAGET'S THEORIES FOR CONTEMPORARY INFANCY RESEARCH AND EDUCATION

Beverly Birns and Mark Golden

The theories of all great thinkers have been applied fruitfully to the practical solutions of human problems, but they have also often been misinterpreted and misapplied. Freud's discoveries concerning psychosexual development and the infantile roots of neurotic behavior have led to the liberalization of childrearing. Extreme permissiveness toward sexual and aggressive behavior in early childhood was not recommended by Freud, although his theories were used to justify such parental practices. Piaget's ideas have also been misinterpreted and misused. Piaget's work contains a profound fund of knowledge about how an individual gains understanding of his world from birth to adolescence. His descriptions of cognitive development during infancy have provided us with an understanding of the processes of development during the preverbal period, which is the particular focus of the present chapter. We shall draw attention to two possible misapplications of his original ideas: (1) the attempt to transform his observations into standardized infant tests; and (2) the current widespread attempt to accelerate sensori-motor development in infants, and particularly in disadvantaged babies.

This chapter will be divided into 6 sections: Piaget's contribution to our understanding of cognitive development in infancy; reasons for the recent acceptance of Piaget's ideas by American psychologists; a description of our own research on social class and cognitive development in infancy; possible misapplications of Piaget's theory to sensori-motor development; recommendations to educators; and certain limitations of Piaget's theory.

Piaget's Contribution to Our Understanding of Cognitive Development in Infancy

In discussing Piaget's theory we must remember that Piaget considers himself to be a genetic epistomologist: He is concerned with how knowledge about the world is acquired by human beings. He is not particularly concerned with individual or group differences, with motivational factors in learning, with the influence of mother-child interaction on learning, or with the educational problems of the socially disadvantaged. Furthermore, he is only secondarily concerned with the application of his ideas to education.

Piaget describes the stages by which a child develops from a primitive reflex organism to an adult who can deal with logic and mathematics. He describes this development in terms of four stages, the names of which refer to their outstanding characteristics. Briefly, the stages are: Stage 1, the sensori-motor stage (approximately the first 24 months of life); Stage 2, the preoperational stage (approximately 2 to 7 years); Stage 3, the concrete operational stage (approximately 7 to 11 years); and Stage 4, the stage of formal or abstract thought (adolescence and adulthood). Piaget's observations on sensori-motor development, the particular focus of this chapter, appear in *Play, Dreams, and Imitation in Childhood* (1962), *The Origins of Intelligence in Children* (1952), and *The Construction of Reality in the Child* (1954).

During the sensori-motor period, according to Piaget, the child acquires knowledge of his world through direct action. For Piaget, the importance of this period for later cognitive development is based on his view that thought is internalized action and that cognitive development is continuous. The acquisition of adaptive sensori-motor schemas or behavior patterns in infancy is the foundation for all later symbolic and abstract thought.

Piaget did not discover new behaviors in infancy that had not been observed by others. This contrasts strongly with his contributions to developmental psychology of older children, many of which were highly original. What Piaget did contribute is a systematic description of the stages or processes by which infants acquire knowledge of spatial relations, object permanence, causality, etc. Many infant tests include such items as visual pursuit of moving objects, which may be observed in infants in the earlier months of life, and the search for hidden objects seen at approximately 9 months of age. Piaget's unique contribution is to link such apparently discrete behaviors and to show how they represent different stages in the development of object permanence. In contrast to the normative approach to infant tests which relates such behaviors to chronological age, Piaget is not concerned with the age at which babies acquire these behaviors. He is concerned with the process or sequence in which new behavior patterns or stages appear, which he believes to be invariant and universal in cognitive development. His original observations on the development of sensori-motor intelligence were based only on his own three children, which is probably one of the reasons why they were not accepted in this country when they were first published. However, the sequence of stages that he described has recently been validated on hundreds of children (Corman and Escalona, 1969; Wachs, Uzgiris, and Hunt, 1967; Gouin-Decaire, 1965).

Beverly Birns and Mark Golden

Reasons for the Recent Acceptance of Piaget's Ideas by American Psychologists

Why have Piaget's ideas become so widely accepted by American psychologists now, after more than 30 years of neglect? The reasons derive both from within psychology and from society's pressures to meet the educational needs of children.

During the first years of the twentieth century, child psychologists had been concerned with either clinical problems or tests and measurements. The clinical theorist most related to Piaget's work was Freud. Although Freud was interested in affective, emotional, and personality factors in contrast to Piaget's emphasis on cognition, Freud was nonetheless the first theorist to focus attention on infancy and early childhood as the basis for later development. Freud's idea of personality development as a continuous process from birth through adolescence was an historical antecedent to Piaget's epigenetic theory of cognition.

The other development logically preceding an interest in Piaget's work was American concern with the "test" approach to intelligence. Early standard intelligence tests for children relied on the normative approach, initiated by Binet and Simon, and further developed in the United States by such workers as Terman and Merrill, and Gesell. Standard intelligence tests provide useful information about the average age at which children are able to perform certain tasks and how individual children compare with their peers. However, they have failed to contribute to our understanding of how intelligence develops. In the past decade American psychologists concerned with intelligence in children have become increasingly interested in understanding the mechanisms or processes underlying intellectual development. Piaget attempts to deal with this issue. For these reasons American psychologists are now more receptive to ideas that Piaget first published many years ago. Piaget's observations on the development of sensorimotor intelligence have been particularly well received for the following reasons.

First, there has been a renewed interest in the infancy period, reflected in the marked increase in the number of studies and publications devoted to infancy during the past decade. Another factor contributing to our national interest in infancy derived from the studies of maternal deprivation during the 1940s and 1950s by Bowlby (1952), Spitz (1945, 1946), Goldfarb (1945a, 1945b), Skeels and Dye (1939), and others. These studies showed that maternal deprivation in infancy led to immediate and long-term deficits in cognitive and personality development. This research provides evidence for the view that infancy is a critical period for cognitive and personality development. A particularly poignant study by Skeels and Dye (1939) described infants in an orphanage who were so retarded that they were transferred to an institution for the retarded and assigned to individual care-taking by adolescent female retardates. Within a few months the infants were functioning at a normal intellectual level, while their counterparts in the orphanage continued to drop in their intellectual performance. The long-term consequences of the kind of maternal deprivation experienced by infants in residential institutions were dramatically documented in a recent 30-year follow-up study by Skeels (1966) which showed vast differences in the social and intellectual functioning in the original two groups of babies.

Second, there has been great disillusionment with infant tests due to their poor ability to predict later intelligence, which has both theoretical and practical implications. When infant tests were first developed, it was expected that performance on these tests could be used to predict later intelligence. This expectation was not borne out. Performance on infant intelligence tests such as the Cattell or Gesell do not correlate highly with later measures of intelligence, except for children with organic impairment who do poorly on both. The following alternative hypotheses for the failure of infant tests to predict later intelligence have been proposed. The first is that there is discontinuity between sensori-motor and verbal intelligence—that the perceptual-motor tasks on infant tests are qualitatively different from the verbal-conceptual tasks of later intelligence measures. An alternative explanation is that there is continuity between the preverbal and verbal levels, but that infant tests do not measure precursors of later cognitive ability. The present authors hoped that the behaviors that Piaget described might be more related to later cognitive development than to standard tests and could serve as the basis for better infant intelligence tests.

Piaget's ideas have also been of great interest to early childhood specialists concerned with the educational problems of preschool disadvantaged children. As a nation we became aroused by the poverty of one-quarter of our youth and in particular by the alarming rate of school failures among poor children which perpetuates their poverty. In addition to the inadequacies of the school system to provide for the special educational needs of poor children, it was also found that many of these children were already far behind their middle-class peers in cognitive development by the time they entered school (Almy, Chittenden, and Miller, 1967; de Meuron and Auerswald, 1969).

Head Start was established to provide compensatory preschool education to prepare poor children to enter elementary school on a more equal basis with middle-class children. Under social pressure to do something about the problem immediately, Head Start programs were hastily conceived and applied on a mass scale. These programs varied tremendously in quality and effectiveness. After initial optimistic and perhaps naïve expectations for Head Start, the recent rash of negative findings from long-term evaluations of these programs has led to pessimism, on the one hand, and false conclusions, on the other. The best programs demonstrated significant gains in intellectual performance during the preschool period. However, these gains seem to disappear by the second or third grade. By this time, Head Start children do not seem to maintain their initial superiority over children who have not had Head Start experience.

These findings have led to two conclusions that we believe are false: (1) Head Start gains are not maintained in elementary school, even in the best programs, because of inadequacies in the preschool program. In regard to this conclusion, it seems more likely that Head Start gains are washed out because of the failure of the public school system to provide excellent education. (2) Head Start has failed because it was too late. Proponents of this view have recommended that compensatory education should begin in infancy. There is an implicit assumption that poor children are maternally or culturally deprived in infancy in a way that is less severe but similar to that experienced by the institutionalized infants described earlier. On the basis of our own research as well as that of other investigators, we feel that there is no evidence of a cognitive deficit in poor children in infancy. Hence the recommendation for cognitive enrichment for such children during the infancy period may be based on a false assumption and therefore may be unwarranted.

The recent growth of the infant day-care center movement has been spurred by a national concern with the increasingly burdensome welfare costs, in particular the Aid to Dependent Children (ADC) program. Day care for infants of welfare families is being provided as a means for reducing the welfare rolls by enabling mothers of infants and young children to obtain training, to work, and to get off welfare. Early childhood educators widely believe that such infant day-care centers should be a downward extension of Head Start. While we believe that our knowledge of child development should be used to provide infants in day care with an optimal environment for cognitive and personality development, we are concerned by the overemphasis on the idea of accelerating sensorimotor development. Our reasons for concern will be discussed in the final sections of this chapter.

Description of Golden and Birns' Research on Social Class and Cognitive Development in Infancy

We had assumed that Piaget's observations on sensori-motor intelligence could serve as the basis for better infant tests, which would predict later intelligence and at the same time show social class differences earlier than standard infant tests. The earliest age at which such differences have been demonstrated is age 3. The direct impetus for our research stemmed from some observations by the junior author of this chapter while he was a member of Dr. Sibylle Escalona's research team. While involved in a validation study of the Piaget Object Scale, he had occasion to do some pilot work on the Object Scale with a small number of impoverished babies under 2 years of age who were living at home with their families. In contrast with the performance of the working-class and middle-class infants he had seen previously, these infants responded more like the apathetic institutionalized babies described by Spitz and others. Either they refused to search for the hidden objects at all or gave up after a few trials. The staff pediatrician who cared for these infants described them in the following terms: "Many of these babies seem so bright and alert during their first few months, but by the time they are about a year old, the light seems to have gone out of their eyes."

On the basis of these observations we decided to do a study

of social class differences in cognitive development in children between 12 and 24 months of age, using the Piaget Object Scale and the Cattell Infant Intelligence Scale. We had assumed that there would be no SES (socioeconomic status) differences on the Cattell, which has been the case in previous studies using standard infant tests. However, we did expect SES differences on the Piaget Object Scale, which seemed to be more related to later cognitive development for the following reasons: Whereas infant tests such as the Cattell assessed children's intellectual development in terms of their increasing perceptual-motor skills, the Object Scale seemed to tap changes in cognition as measured by the child's increasing knowledge about objects.

In the modified version of the Escalona-Corman Object Scale, the child's instrumental response remains the same: He is required to search for objects hidden under a cloth under increasingly complex conditions. At first, he is required to find an object hidden under a single cloth. His ability to do so reflects his understanding that objects exist when they are no longer in the perceptual field. Then he is required to search for the object under the correct cloth, when there is more than one cloth in the field. His ability to search only under the correct cloth, the one where he sees the object disappear, reflects his increasing understanding of the laws of spatial displacement of objects. If an object is hidden under cloth *B*, it cannot miraculously appear under cloth *A*, where he may have found it previously. Still later the child comes to understand the laws of invisible displacements of objects. The examiner hides an object in his hand, places his hand under a cloth, and reveals his empty hand to the child. If the child then searches under the cloth, he has inferred that there has been an invisible displacement of the object, i.e., that the examiner must have left it under the cloth.

We did a cross-sectional study (Golden and Birns, 1968) in

which 192 black children of 12, 18, and 24 months of age, from different SES groups, were compared on the Cattell Infant Intelligence Scale and the Piaget Object Scale. Children from the following three SES groups were studied: (1) Welfare Families—neither mother nor father currently employed, family on welfare; (2) Low-Educational Occupational-Status Families—neither mother nor father has more than a high school education, or has been employed at more than unskilled or semiskilled jobs; and (3) Higher-Educational Occupational-Status Families—either mother or father has some schooling beyond high school, or has been employed at skilled or professional jobs.

The results of this study confirmed the previous research findings by other investigators. Social class differences in intellectual development were not detected during the first 2 years of life. Comparing children from fatherless welfare families with children from stable low-income and middle-income families, using two relatively independent measures of cognitive development, and making every effort to overcome motivational factors that seriously interfere with test performance, we did not find any differences among the three SES groups at 12, 18, or 24 months of age on either the Object Scale or the Cattell.

Although many of our group of welfare babies lived in abject poverty, the possibility remained that a sample of mothers who would participate in this kind of study were unlike those who refused to participate. If this were true, at 3 years of age, when one would expect to find SES differences, these very poor children might be indistinguishable from their economically advantaged "peers." Therefore, we did a follow-up study of the 18- and 24-month-old children who were tested with the Stanford-Binet at 3 years of age (Golden et al., 1971). Whereas there were no social class differences at 18 and 24 months of age on either the Cattell or the Piaget Object Scale, when the same children were retested on the Binet at age 3 there was a highly significant 23-point Mean IQ difference between our highest and lowest SES groups.

Although both the 24-month Cattel ($r = .60$, $p < .005$) and the Object Scale ($r = .24$, $p < .05$) were significantly correlated with the Stanford-Binet at 3 years, neither of these measures at 18 months correlated significantly with the 3-year Stanford-Binet.

In discussing the results of this research we concluded that the sensori-motor and verbal levels of intellectual functioning were qualitatively different. In order to acquire sensori-motor skills in New York City, rural America, Geneva, or Nairobi, a baby needs an opportunity to acquire knowledge of his environment through his own explorations and experience with that environment. All babies learn that objects appear and disappear, that if a container is tipped over, its contents spill out, etc. Little of children's knowledge about the world at this age is socially transmitted to them by their parents. This does not imply that maternal behavior does not affect development in infancy. Studies by Blank (1964), Rubenstein (1967), and Yarrow (1971) all suggest a relationship between maternal behavior and infant behavior. What we are suggesting, however, is that precocity during the first 18 months is unrelated to later cognitive ability. After 18 months of age, when language enters the picture, more and more of what children learn about the world is taught to them by their parents. And it is at this time that the parent's education becomes important. It is at this time that socioeconomic and educational differences in the family begin to have their impact on children's intellectual development.

If parents answer questions and give explanations, if past and future experiences are discussed and described, the child acquires a cognitive mode of interaction that prepares him to adapt to a middle-class classroom. However, if this kind of verbal interaction is discouraged, the child may retain an action-oriented mode of problem-solving, which will interfere with school performance.

Possible Misapplications of Piaget's Theory to Sensori-Motor Development

In the present section we shall discuss what we believe to be two possible misapplications of Piaget's ideas on sensori-motor development: (1) The use of Piaget-type scales as newer and better infant IQ tests, and (2) the use of his concepts to justify the acceleration of sensori-motor development by specific teaching techniques.

In regard to the Piaget scales developed by Escalona and Corman and those by Wachs, Uzgiris, and Hunt, these studies proved invaluable by validating on hundreds of children Piaget's original observations made on his own 3 children. However, the Escalona-Corman scales were never intended by the authors to be used as standard intelligence tests, but rather as research instruments in their study of cognitive development in infancy. Further, the assumption that these measures are more predictive of later intelligence than standard infant tests are has not been borne out by our research (Birns and Golden, 1972).

In regard to attempts to accelerate sensori-motor development, one should re-examine Piaget's views about how sensori-motor development occurs. According to Piaget, during the first 2 years of life the baby gains knowledge of his world by his own actions and not by being taught by his parents. He learns to follow objects as they move through space, whether the objects are mobiles, his own hands, the comings and goings of other people, or cockroaches crawling across the floor. He practices following moving objects. He learns to coordinate eyes and hands to reach for desired objects. Whether the object is a doll hidden under a napkin by his father or a cookie that has previously been put away in the closet by his mother, the child learns that these objects exist outside his field of vision. Babies learn object permanence through every-day experiences, whether or not their mothers play hiding games with them.

Given an average expectable environment, children will learn object permanence. However, after 18 months of age, when more and more of what a child learns is acquired from his parents through language, the role of direct teaching by parents plays a significant role in the child's cognitive development.

We are particularly concerned with the current widespread application of Piaget's ideas in accelerating sensori-motor development in day-care centers for disadvantaged infants. In the first place, we are not convinced that they need compensatory enrichment, since no deficit in cognitive development has been demonstrated in these infants during this early period. Secondly, we consider it unsound to measure the value of infant day programs on the basis of their success or failure to accelerate children's cognitive development during the first 18 to 24 months. In assessing the social utility of infant day-care programs, we feel that more emphasis should be placed on the impact of such programs on the family, in making it possible for mothers to work and to break out of the cycle of poverty. Further, of major importance is the fact that no one has demonstrated the assumption that accelerating or enriching sensorimotor development will have any effect on later cognitive functioning.

Recommendations to Educators

Professionals and nonprofessionals who plan to work with infants in nurseries or day-care centers should become thoroughly familiar with Piaget's writings on the sensori-motor period. They should become familiar with Piaget's ideas, not to teach infants such concepts as object permanence, spatial relations, causality, etc., but to gain a deeper understanding of how infants, through their own explorations and activities, acquire knowledge about the world. Through such understanding perhaps they can develop a respect for the infant as a scientist-explorer on his voyage of discovery through his new and unchartered world. Child-care workers can get much pleasure from simply watching or paying attention to the infants in their care, without feeling that they must do something to speed up the child's intellectual development.

Paying attention to infants may well turn out to be the most important thing which adults can do to stimulate development during the infancy period. This does not mean that persons in contact with babies should not interact with or play with infants or lavish affection on them. However, along with affection, babies also need the opportunity to explore and to master their environment. Studies of institutionalized infants suggest that infants who are merely provided with routine care, with little other attention from adults, are apathetic and seem unmotivated to explore the environment.

In addition to paying attention to infants, can child-care workers use their knowledge of Piaget to foster babies' development or learning? Piaget's ideas about the process by which infants acquire new behavior patterns may be useful. Piaget believes that children's adaptive intelligent behavior is guided by two complementary processes, which he refers to as "assimilation" and "accommodation." Assimilation is simply the application of an established behavior pattern to a familiar or new situation. If the behavior is successful, the child is not forced to change his behavior in the new situation. However, if the behavior is not successful, the child must adapt or change his behavior to the new situation. Accommodation is changing an existing behavior pattern that does not work in the new situation.

Piaget believes that a child is more likely to accommodate his behavior to solve a problem when the new behavior that is required differs *only slightly* from those already in his repertoire. If parents, day-care workers, and other adults caring for infants understand what the infant is trying to do, they may be able to facilitate learning by providing just the right environmental input or match between what the child is able to do and what he cannot yet do. For example, the baby may be reaching for something that is far beyond his grasp. After several unsuccessful attempts he cries. The care-taker who has not been paying attention to the baby may pick him up to comfort him or try to stop the crying by introducing a new toy or activity, without taking into account what the baby is trying to do; or the adult may simply pick up the distant toy and hand it to the baby; or even worse, the adult may ignore the infant in order not to "spoil" the child by paying too much attention. The person who understands what the child is trying to do can help by placing the toy closer but still out of reach, thereby enabling the infant to get the toy himself. In this way the infant's development is facilitated without direct teaching.

Certain Limitations of Piaget's Theory

Although scientists are obviously justified in setting the boundaries of their field of inquiry, those seeking practical applications of the theory ought to consider any one particular theory within the context of the total field that is their concern. Thus, with full appreciation of Piaget's contribution to our knowledge and understanding of the development of thought in childhood, we ought not to ignore motivational and emotional aspects of development.

Not only do infants develop in a world of objects that they explore, touch, taste, throw, move, and act upon in an infinite variety of ways, but they also grow in a world of people. It is our contention, backed up by considerable research, that the care-taking adults, through their attentiveness, affection, enthusiasm, and verbal and social interaction, provide the base for the infants' learning about the world—not only its objects and physical properties, but its people including themselves.

Consider Erikson's concepts of basic trust, autonomy, and initiative. It is during infancy that the initial feelings about the self and others develop. Is the world a safe, responsive, and exciting place? Or is it full of prohibitions and unfulfilled needs? Are people helpful and encouraging or do they only limit and coerce? How does the developing child feel about himself and what can he learn and master? The 2 year old may not articulate these questions, but neither does he ask under which cover is the hidden object. Nevertheless, this may be the most important area of early learning.

One further departure from Piagetian theory concerns the role of language in cognitive development. In *Play, Dreams, and Imitation in Childhood* (1962), Piaget propounds the very novel and provocative view that language is based on the prior development of the symbolic function, which initially is nonverbal. Although he describes the process by which children acquire language, he relegates it to a secondary role in thinking. While we have expressed the view that most of what babies learn during the sensori-motor period is acquired through their own activities, once language enters the picture care-taking adults can then begin to play a more active role in teaching the child about the world through the use of language.

We have postulated that sensori-motor and later cognitive development may be discontinuous and that children at different ages might well require different kinds of stimulation. We should finally emphasize that at all ages babies and children need warmth, appreciation, and recognition for newly acquired behaviors as well as respect for individual differences in temperament, personality, and styles of learning.

REFERENCES

Almy, M., Chittenden, E., and Miller, P. *Young children's thinking*. New York: Teachers College Press, Columbia University, 1967.

Birns, B., and Golden, M. Prediction of intellectual performance at 3 years from infant tests and personality measures, *Merrill-Palmer Quarterly 18*, 1 (1972), 53–58.

Blank, M. Some maternal influences on infants' rates of sensorimotor development, *Journal of the American Academy of Child Psychiatrists 3* (1964), 668–687.

Bowlby, J. *Mental care and mental health*. Geneva, Switzerland: World Health Organization, 1952.

Corman, H. H., and Escalona, S. K. Stages of sensorimotor development: A replication study, *Merrill-Palmer Quarterly 15*, 4 (1969), 351–361.

De Meuron, M., and Auerswald, H. Cognition and social adaptation, *American Journal of Orthopsychiatry 39*, 1 (1969), 57–67.

Golden, M., and Birns, B. Social class and cognitive development in infancy, *Merrill-Palmer Quarterly 14* (1968), 139–149.

Golden, M., et al. Social class differentiation in cognitive development among black preschool children, *Child Development 42* (1971), 37–45.

Goldfarb, W. Psychological privation in infancy and subsequent adjustment, *American Journal of Orthopsychiatry 15* (1945a), 247–255.

Goldfarb, W. Effects of psychological deprivation in infancy and subsequent stimulation, *American Journal of Psychiatry 102* (1945b), 18–33.

Gouin-Decarie, T. G. *Intelligence and affectivity in early childhood*. New York: International Universities Press, 1965.

Piaget, J. *The origins of intelligence in children*. New York: International Universities Press, 1952; London: Routledge and Kegan Paul, 1953.

Piaget, J. *The construction of reality in the child*. New York: Basic Books, 1954; London: Routledge and Kegan Paul, 1955.

Piaget, J. *Play, dreams, and imitation in childhood*. New York: Norton, 1962; London: Routledge and Kegan Paul, 1951.

Rubenstein, J. Maternal attentiveness and subsequent exploratory behavior in the infant, *Child Development 38* (1967), 1089–1100.

Skeels, H. M. Adult status of children with contrasting early life experiences, *Monographs Society for Research in Child Development 31*, 3 (1966), 1–65.

Skeels, H. M., and Dye, H. B. A study of the effects of different stimulation on mentally retarded children, *Proc. of the American Association on Mental Deficiencies* 44 (1939), 114–136.

Spitz, R. A. Hospitalism: An inquiry into the genesis of psychiatric conditions in early childhood. In O. Renichel et al. (eds.), *Psychoanalytic study of the child*. New York: International Universities Press, 1945. Vol. 1, pp. 53–74.

Spitz, R. A. Hospitalism: Follow-up report. In O. Renichel et al. (eds.), *Psychoanalytic study of the child*. New York: International Universities Press, 1946. Vol. 2, pp. 113–117.

Wachs, T. D., Uzgiris, I., and Hunt, J. McV. Cognitive development in infants in different age levels and from different environmental backgrounds. Paper presented at the Biennial Meeting of Society for Research in Child Development, New York, 1967.

Yarrow, L. J., Rubenstein, J. L., and Pederson, F. A. Dimensions of early stimuli differential effects on infant development. Paper presented at the Biennial Meeting of Society for Research in Child Development, Minnesota, 1971.

CHAPTER 6

LANGUAGE AND THOUGHT

Eleanor Duckworth

The title of this chapter, as many readers will recognize, is an affectionate tribute to Piaget's first book, *The Language and Thought of the Child* (1926). It is also an acknowledgment of the fact that, in all that has been written about Piaget in the last few years, very little has been said about his views of the role of language in intellectual development. This is not too surprising, because he has concerned himself very little with language and has written very little about it. But it is worthwhile to look at his views on the role of language and to see why he is not more concerned with it than he is.[1]

The title of his first book, mentioned above, and section titles such as "Grammar and Logic" in his second (*Judgment and Reasoning in the Child*, 1928) would seem to belie the statement that he has not been concerned with language. But these two are among his five "early books," which are quite different from all his work since then in style of investigation and even, to some extent, in theoretical orientation. He himself now finds them of little interest. Yet even then his approach was original enough and he still subscribes to the main thesis. He did not assume that language and linguistic forms that children use coincide with their thinking. He was trying to look behind the language to the thoughts of which the language may or may not have been an adequate expression. Where language was inadequate, he did not assume that there was a direct parallel with the way that the thinking was inadequate. Rather, he took the language inadequacy as an indication of something to look for and proceeded to look for it. Conversely, where language seemed to be adequate, he did not accept it at face value, but, once again, tried to go beyond the language to see whether it meant what it seemed to mean.

What Piaget now finds less interesting in these books is that his insights into what children really were capable of doing intellectually were nonetheless based on what they *said*. In his later work (after 1935) he looked much less at what they said and looked instead at what they did. Through watching the development of sensori-motor intelligence, before the development of language in a small child, he found that the roots of logic are in actions and not in words. He followed the development of this logic of actions through to adolescence, finding at every step of the way that children were able to carry out activities that demand a good deal of intelligence without necessarily using language that reveals this. In sum, his early insight was that language often is a misleading indicator of the level of a child's understanding; a second insight was that there is a good deal of logic in children's actions which their verbal formulations simply do not reveal.

In the next sections, I shall use a number of different words in referring to what is behind language: meaning, sense, knowledge, ideas, logic, awareness, feeling, and others. I do not assume that these are all the same thing, but for our purposes we need not define distinctions among them. What they have in common is that they are what is in our heads, if you will accept that image, as opposed to what we make explicit through our language.

Language and Thinking

Even a casual observation of ourselves can indicate that each of us, day by day, has at least some thoughts for which we have to look for words, rather than having only those thoughts that come readymade in words. "I have a feeling that Jack is in that mood again tonight." "What mood?" "Oh well, it's difficult to say." But the speaker surely has an idea of what he means by "that mood"—how he recognizes it, how Jack is likely to act, and how he should act in turn for best results. He might even be able to make those thoughts explicit. But he has them "in mind" even if he never happens to say them.

Again, when we quite easily make thoughts explicit, we might do so in any one of a great variety of ways. In these cases, too, it seems to me that the notion of what we want to say exists in some way, and we then find words with which to say it: "That one is much heavier than mine," "That one is much heavier than the one I have," "The one I have isn't nearly as heavy as that one," or "The one I have is much lighter than that one," and so on. It seems unlikely that it is in formulating the sentence that the speaker becomes aware of the thought. This is not to deny that there are differences in emphasis among these sentences. A person may be aware of the different possible emphases and form his sentence accordingly—in which case that awareness was there before the sentence was formed; or he may not be sensitive to the emphasis, and, if someone were to point out that the way he said it suggested such and such an emphasis, he would be able to say whether he had intended that or not—in which case, once again, his intention was somehow clear to him independent of the words he used to express himself.

On the other hand, people sometimes use language that goes far beyond their thoughts. Some people can dazzle us with elaborate words when they do not really know what they are talking about. Here is a sentence that I have just produced, with no thought at all in my mind: "The implication of the prognosis indicates that the thrust of our ultimate endeavors obliges us to reconsider all our assumptions and to justify the utilization of the heretofore unconsidered media." As I look at it now, I can find some meaning for most—though not all—of it. But I am working hard to *create* the meaning—it certainly is not given to me by the words. Some people seem to talk this way a lot of the time, with very little thought behind their words.

The poetic aspect of language might be said to share something with each of the two preceding situations. As in the first, when the writer starts, he may have little idea of the form that his words will take. Once he has written, he may look at his writing as a critic, to see whether the emphasis he has given does correspond to his feeling. As in the second, a poet's facility with words sometimes leads him to put them together in ways that go beyond any meaning of which he was aware. Some readers may be able to create for themselves, as they read those words, a meaning of which the writer himself remains unaware.

At other times, words can be so poorly chosen that they mask what is really a good idea. If we think it may be worthwhile, we take time to reflect on a muddled sentence and we sometimes find out that these muddled words do in fact point to an interesting idea, which we can rephrase for ourselves if we wish. This rephrasing corresponds to the real nature of the job of an interpreter. When he is interpreting from one language to another—even simultaneously—he does not do a mechanical input-output job. He listens to a big enough piece to grasp the sense of what is said and then puts that sense into words again, in a different language. He necessarily does a poor job if he is translating a topic he does not understand, even if he has boned up on the vocabulary.

One might say, unkindly, that Piaget has a vested interest in discrediting the view that ideas are only as good as the way they are expressed, because anyone who tries to read him firsthand is quite taken aback by his lack of eloquence and even clarity. Yet what is on his mind—if we get beyond the way he expresses himself—is surely significant.

Constructing What We Know

In order to know something or to think about something, then, we do not have to use words. The question for teachers, of course, is how thoughts get into people's heads (or wherever they are) in the first place. Some people believe that, even if we do not always need words to think the thoughts once they are there, it was through words that they got there to begin with. Sometimes, of course, this is true—sometimes we can immediately connect something we are told to what we know already, and the thought becomes our own.

But notice that if we really understand what we have been told, we make new connections *for ourselves*. We are now the master of these new connections and can express them our own way.

If we cannot make these new connections for ourselves, we do not really grasp what we have been told. In fact this is where we are most likely to hold on to the exact words! If a child is told that water runs downhill, he is much more likely to be able to repeat those same words than he is to be able to *rephrase* them with all the meaning that they represent. He is very unlikely, on the basis of a sentence like that, to be able to draw the significant connections—as, for instance, that the outlet from the Great Lakes must be uphill from Quebec City. Piaget's emphasis is that we have to do the work ourselves, making the connections, even if people take pains to point out to us connections that they have been able to make.

It is worth looking at two of the most important of Piaget's books, to see what he has to say about the beginning of our thoughts. *The Origins of Intelligence in Children* (1952) and *The Construction of Reality in the Child* (1954) deal with the development of intelligent behavior in infants before the development of language. The very fact that he has written two volumes on this topic is an important indication of his position. Infants begin to act in intelligent ways from the day of their birth, that is, they make connections, seek consistencies, and modify their behavior in terms of their situation. One might say that they are armed initially with nothing more than their reflexes; a guiding *motivational* rule that might be stated like this, "If I can do it, I will;" and a guiding *intellectual* rule that might be stated, "All else being equal, things will turn out the same." With this equipment, infants proceed to carry out their reflexes in all possible situations, to modify them as necessary, and to figure out which actions in certain situations give rise to more possible things to do. The more they try to do what they can, the better they are able to adapt to the circumstances and the more they start to differentiate and to coordinate what they can do. In this way they come to respond to more aspects of the situation at once. They are making refinements and connections in their actions—"thinking" in their actions—long before there is any use of language.

In their actions, they are constructing what they know about the world, and they are constructing their logic, such as classifying, ordering, conserving. Some things can be grasped, some cannot; some move when they are pushed, some do not; some things fall when you let go of them, others do not. In order to grasp an object, you have to open your fist, put it on the object, and then close your fist—any other order will not work. If a thing swings when you hit it once, it will probably swing when you hit it again. All these insights are independent of language.

What each of them represents is what Piaget calls a scheme. The totality of your schemes is the totality of what you know. At the prelanguage, presymbolic level, your schemes are what you know how to do.

It is important to realize that what is known, a scheme, is more general than any one instance of carrying it out. Let us say that an infant is reaching for an object that is out of his reach, but is sitting on a blanket; he manages to get the object by pulling the blanket toward him (a sophisticated connection to make). And let us suppose that, even if that first time was by accident, he realizes that it was a good procedure, and uses it another time on purpose. Now we can say this is a scheme. But the next time he does it, he will not use *exactly* the same movements. He might reach higher or lower, nearer on the blanket or farther. It might be a coat this time, instead of a blanket, so he would grasp it differently. He might even use the other hand. So it is not simply a motor conditioning. The scheme which is now at his disposition is not simply a matter of superficial movements. It is more like a rule of grammar. Speakers create all sorts of sentences they have never heard, nor even produced themselves before. They have understood the rule of grammar and use it in all sorts of different superficial manifestations. Similarly, once the infant has made the connections, a scheme is at his disposition, and he uses it in all sorts of different superficial manifestations.

An infant does not represent this knowledge to himself. Lying in his crib, he does not recreate in his mind's eye the actions that he might be capable of carrying out. It is when a possibility presents itself to him that he proceeds to act appropriately. Not only has he no language, then; neither has he any other form of representing what is not immediately present.

Piaget describes the development of this ability to represent something by something else in his book *Play, Dreams, and Imitation in Childhood* (1962). The English translation of his title does not describe its contents well. The book really deals with the development of the child's representational ability. Play, dreams, and imitation are three ways children have of representing what they know of the world—language, of course, being another. In this book, Piaget develops the thesis that children's first internal representation—their first "thinking without actions"—takes the form of internalized imitation.

Imitation, too, has its beginnings in the very first days of life, in primitive forms like circular reflexes, imitation of oneself, or involuntary imitation of a sound. But it is relatively late that imitative behavior takes on a really representative function, in the sense that the child imitates something on purpose, making connections that he has never made before and using this imitation to stand for something that he knows is different from his own action. Through imitating something a child develops the schemes by which to understand it—very much as Marcel Marceau comes to understand what is involved in tugging on a rope by miming the action. Piaget describes a case where his daughter, at 1 year and 3 months, catches the feet of a toy clown in her clothing. After disentangling it, she hooks her finger into her clothing in the same way and pulls at it in an effort to understand what happened. She is imitating the feet of the clown with her finger; in other words, her finger is representing the feet of the clown.

Much analyzing and coordinating are involved in imitation. Because the child's own body is the first thing available to him to reproduce with, he acts out his analyses and coordinations. Gradually, he becomes able to internalize these imitations—to carry them out mentally not "in his mind's eye," but "in his mind's body," as it were.

At this age an infant also realizes that things have a continuing existence even when they are not present in his own perceptual field. He realizes that it is not his looking or holding or chasing that brings them into being. He now has a way of evoking things for himself, even when he is not in their presence.

This is the age, too, as all parents know, when a child starts to use language. Words have always been around him, as part of the situations he has been in. But now that he can use one thing to stand for another, words become available to him as a useful aspect of the situation to imitate, as a short-cut evocation of it. Thus, language comes into being as one of several ways the child develops of representing one thing by another.

Now it should be clear in this case that words do not create the sense or intention or feeling; the words accompany it or seek to express it, or to refer to it. At this stage the adequacy or inadequacy of the words is not even a close reflection of the adequacy of the sense that is in the child's mind.

It should also be clear that since words are conventions that do not resemble what they stand for, there is no way that a child who is beginning to tune in to language can be sure what a given word or phrase or sentence refers to when he hears it. He is likely to believe that it refers to what his mind is on at the moment. If he is looking out the window when the neighbor's dog goes by and his mother says "doggie," he might think "doggie" means something going by outside the window; something that lives in the house next door; anything white and brown; or anything running along the street—depending on what he happens to be noticing.

This is the essence of what Piaget has called "egocentric thought"—what is foremost in the child's thoughts is what he believes others are attending to, also.

As children grow up, they realize that what they are paying attention to is not necessarily what other people are paying attention to. But this is a hard lesson to learn, and nobody ever learns it totally. All of us tend to be absorbed with thoughts that we already have, or have once had; if people start to talk generally in that direction, we tend to interpret what they say as things that we have been thinking. The phenomenon isn't restricted to single words. Whole sentences and paragraphs are misconstrued. We are so intent on what we are thinking that we do not realize that the other person is saying something different. Each of us is familiar with conversations that at some point start to run at cross purposes when the participants have different things in mind. They get straightened out (if indeed they do), with exclamations such as "Oh I thought you meant—." Often we are so sure of what the other speaker must be saying that we don't hear the clues to the contrary. "Oh, every time you said Chicago, I was thinking Toronto!"

Small children happen to be particularly attached to what they are focused on at the moment. They are very prone to taking in whatever is said and fitting it to what they thought already.

In *The Language and Thought of the Child* (1926), Piaget notes the extent of this phenomenon in nursery-school-aged children and describes situations he set up to look at it more closely. For instance, he would explain to one child how a water tap works. He ascertained how much the child understood. Then that child proceeded to explain it to another child. In general, what he found was that the children explained as if they assumed the other child knew already. The language they used was more in the nature of very sketchy reminders, pointing out a number of highlights and assuming that the connections involved were clear. The child clearly tries to explain (the explanations are full of "You see?" and "And then it comes here, see?"), but he talks as if what is obvious to him as he now is engaged in his explanation is—by that very fact—obvious to everyone else.

Even more impressive is the fact that the child to whom this explanation is being given appears to have equal confidence! He constructs his own notion of how the tap works and he interprets the words of the other child so that they fit into what he is thinking! He is sure he understands everything the other child is trying to tell him. Thus, with these two children—one with one view of how a tap works, the other with his own view—lots of words were passed between them, but the words made little headway in conveying any meaning from one child to the other.

There is a clear message for teachers here. Words that people hear—and the younger the child is, the stronger the case—are taken into some thoughts that are already in their minds, and those thoughts may not be the ones the speaker has in mind. A good explainer can anticipate what our interpretations are likely to be—on the basis of what we are likely to know, to be thinking about, or to have noticed already—and can fit the explanation to that, with phrases like, "I don't mean *X*, I mean *Y*," or "Look, did you notice this?"

There is a message for listeners, too. A good listener, or a good understander of explanations, is aware that his first interpretation of what is being said may not be the right one, and he keeps making guesses about what other interpretations are possible. This ability is singularly undeveloped in little children but it should be highly developed in good teachers, who try to listen to what children are trying to say to them.

So far we have been dealing with knowledge of how the world works. In summary, each individual has to construct his own knowledge. Sometimes we can be helped by what other people tell us, but we still have to do the work ourselves. Often, we can say things we have been told without understanding what we say.

Logic Is Deeper Than Language

Another aspect of the relation between language and thinking needs to be considered. Language is full of expressions of logical relationships. Some people believe that teaching children careful use of these linguistic forms helps to develop clear ways of thinking. Language reflects levels of classification and subclasses—for instance, some birds are sparrows and all sparrows are birds. It reflects cross-classification—peas are seeds and peas are food, at one and the same time. Language reflects ordering relationships—father, niece, grandfather; big, biggest; more, less. It reflects logical connections—because, even though, if . . . then.

Since these relationships are all built into language, one might think that children have only to pay attention when they are used in order to understand the relationships. But this is equivalent to assuming that, when a child hears "doggie" as he looks out the window, he knows exactly what "doggie" refers to. On the contrary, the evidence indicates that in this case, also, children think their own thoughts and interpret the language in their own way, which does not necessarily correspond to what the speakers around them have in mind.

One of Piaget's very first questions, in a 1921 article that predated even the first of his books, dealt with the way children use "some" and "all." Later he looked at this question again, when his theoretical position resembled what it is now. He found that, although children had a certain familiarity with what "some" and "all" meant, their understanding of these terms was not what we usually understand by them. In one investigation, children had a collection of red and blue circles, and red squares (no blue squares). One of the questions they were asked was, "Are all the squares red?" Five and six year olds would generally answer, "No, because there are some red circles, too."

In another investigation, he found that, although children agreed that horses and cows were both animals, they would look at a collection of 6 horses and 2 cows and say that there were more horses than animals, because there were only 2 cows.

In both these cases, the children clearly have some relationships in mind other than the ones that these questions suggest to us.

One of Piaget's collaborators, Christofedes-Papert (unpublished study), repeating this second experiment, asked an additional question. After agreeing that there were 6 horses and 2 cows and that they are all animals, the children were asked, "Are there more horses here, or more animals?" "More horses, because there's only two cows." This time the experimenter asked, "What did I ask you?" "You asked me if there are more horses or more cows." "Ah, alright. Now listen to this question. Are there more horses or more animals?" "There are more horses, because there's only two cows." "What did I ask you?" "If there's more horses or more cows," and so on.

The child is making his own comparison between horses and cows. In spite of the experimenter's words, that is the comparison he continues to make. Piaget's interpretation is that he cannot think of horses as both horses and animals at the same time. Whatever the case, the experimenter's words are not getting through to the child. The words themselves can't put their own meaning into his head.

The same phenomenon is in evidence in other transformations children make in rephrasing things they are told or asked. Sinclair (1967) asked children to give more candies to one doll than to another, which they did, and then asked what she had told them to do. The children said, "Give a lot to that one and a little to that one," which is not necessarily the same meaning but which indicates how the children understood "more" and "less."

This is reminiscent of work done by linguists such as Labov (1970) in studying dialect differences in ghetto children's speech. If the children were asked to repeat a sentence of a form that did not correspond to their grammar (for instance, "I asked Alvin whether he knows how to play basketball"), they repeated the sentence, but with their own grammar ("I asked Alvin do he know how to play basketball"). It was not the words they retained, it was the sense. Then the sense was translated back into words, words that said the same thing but were not the same words.

There is a difference between these children and the children in the Geneva studies. These children do understand exactly the sense of the words they have heard, as intended by the speaker. They have assimilated that sense accurately and have re-expressed it in their own words. The children in the Geneva studies assimilate their own sense, not the speaker's intention, and their rephrasing reveals their own sense. But the similarity between the studies is that the words are so far from being important that they are instantly forgotten and replaced by words that more adequately—from the point of view of the child—express the sense of what the child understood.

Another early concern of Piaget's was children's use of conjunctions that express logical relationships. In *Judgment and Reasoning in the Child* (1928) he showed that, even though children could use these conjunctions grammatically, the logic involved in their use was far from clear. He proposed that children sense some sort of relationship between two propositions and sense that there are certain kinds of words that can be used to indicate such relationships. But the relationships are not clear, and the conjunctions, as a consequence, get all mixed up. He observed this in children's spontaneous use of these conjunctions, so he gave children sentences to complete and found completions like these:

Peter went away even though . . . he went to the country. (6 years.)

It's not yet night even though . . . it's still day. (8 years.)

The man fell from his bicycle because . . . he broke his arm. (7 years.)

Fernand lost his pen, so . . . he gave it to a kid and he lost it in the park. (4 years.)

Fernand lost his pen, so . . . he found it again. (7 years.)

I walked an hour more even though . . . I like walking. (6 years.)

I did an errand yesterday because . . . I went on my bike. (6 years.)

Other research that Piaget points to as supporting his view of the relationship of language and logic is that of Oleron (1957) and of Furth (1966). The general gist of their findings is that deaf children develop the same logical abilities as hearing children, at just about the same ages, without the contribution of a constant verbal bath from wiser adults.

Piaget has speculated, and there is some elementary evidence to support him, that blind children would be more impeded than deaf children in developing these same notions, since the roots of logic are to be found in the coordination of actions, and this coordination is much facilitated by being able to see how actions are coordinated. The infant's hand-eye coordination is one example.

A large-scale and subtle study by Sinclair (1967) also looked at relationships between language level and intellectual level. Her expectation at the outset was that children who did better in Piaget situations would have a better language level, and she hypothesized that this more sophisticated language would be what accounted for their more sophisticated thinking. The two Piaget situations she studied were conservation (the classic pouring of liquids experiment) and serial ordering (putting ten sticks in order of length.) She did, in fact, find that children who succeeded in these tasks had more sophisticated language abilities, in a number of different ways. But then she proceeded to teach the less advanced children the language of the more advanced, believing that this would help them in their cognitive tasks. She found, first of all, that it was extremely difficult to teach them the language patterns. Their language seemed to be limited by their level of understanding. Nonetheless, she did make some headway with most of the children. Yet she found that, even with more sophisticated language, they did not, on the whole, do better than they had before on the intellectual tasks. Contrary to her original hypothesis, she concluded that language development is dependent on the level of thinking rather than being responsible for the level of thinking.

The pedagogical implications here seem to be fairly clear-cut: teaching linguistic formulas is not likely to lead to clear logical thinking; it is by thinking that people get better at thinking. If the logic is there, a person will be able to find words adequate to represent it. If it is not there, having the words will not help.

This point may be clarified by referring once again to the development of intelligence at a preverbal age. Earlier in this chapter we considered the totality of one's schemes to be the totality of what one knows. For an infant, schemes are what he knows how to do. For an older child or for an adult, schemes are, in addition, the thoughts in his head, his ways of putting things together and making sense of the world. Coordinating thoughts is as big a job as coordinating actions in an intelligent way. Think of trying to develop in a baby the scheme of pulling the blanket to get the toy that is on it, by holding his arms and maneuvering him through it. If he is just about ready to make the connection for himself, he may "get it" when he is shown in this way. But, if he doesn't see the connection, it won't become his scheme. The next time around, it still will not be part of his repertoire. Similarly, an older child will not be able to internalize and to make his own some logical connection that somebody makes for him—even many times—if that connection is far beyond him.

Drilling children in sentences of the "If . . . then" format is not likely to develop in them the notion of logical implication, contrary to Bereiter and Engelmann's (1966) expectations. Some of Piaget's collaborators gave children a collection of different sorts of dolls and asked them to make subgroups corresponding to certain descriptions. One of the descriptions was, "If it is red, then it is big," and for this the children would take only red, big dolls. They "heard" those words to mean, "It is red and it is big," and they would accept no nonred dolls in the collection, even if the experimenter suggested it. It is doubtful that Bereiter and Engelmann's preschoolers would do any differently, despite their drillings.

Building New Connections

Thoughts are our way of connecting things up for ourselves. If somebody else tells us about the connections he has made, we can only understand him to the extent that we do the work of making these connections ourselves. Making connections must be a personal elaboration, and sometimes a person is simply not capable of making the connections that someone is trying to point out to him.

Often, of course, we do use words in our thinking, especially if we are trying to elaborate something new. Through trying to make things explicit for ourselves, we can see our own loose ends and we can see where we must make still other connections. Similarly, words from somebody else can point out loose ends in the way we have put our thoughts together. They can also point to connections that someone else has made, so that we may be able to make the same connections when the way has been suggested. This they do as any other representation does. A Calder wire sculpture can draw our attention to the soulful fatness of a cow. A museum water-flow model can indicate something about river currents. A film can make us stop and reconsider some of our beliefs about people. A ballet can lead us to new insights into the nature of madness. Painting, pantomime, model-building, mathematical symbols—to the extent that these have at least some aspect of recreating, reconstructing, or representing something other than what they themselves are—represent, for the person who creates them, something of his understanding. It is no clearer in these cases than in the case of language that what they evoke in the beholder is the same understanding or knowledge or feeling as that which the producer had in mind. They call upon the beholder to make his own connections, in order for the representation to make sense to him.

In passing, notice how Piaget's very methodology reflects his views of language. Piaget and his researchers engage in a rather loose discussion with a child. The researcher has a number of key questions in mind, to be brought up in a standard order. But the phrasing of the questions and ensuing discussions with the child depend on the child's reactions. Piaget is criticized by many psychologists for not having a standardized format—a fixed set of questions, phrased in a fixed way, so that exactly the same words are used with each child. The point of this standardization is to guarantee that each child is dealt with in the same way. But from Piaget's point of view, standardizing the words has little to do with standardizing the problem for the children. The words are only a way to get the thinking going. There is no guarantee that the same words will cue in the same way for every child. It is important to vary the words used until they make contact with the child's thinking. Reaching the child is what has to be standard. Sticking rigidly to a fixed formula can almost guarantee a *lack* of standardization.

Learning to Spell

The argument that we have been advancing in this chapter so far is that there is no need to give children "language tools" in order to facilitate clear thinking, intelligence, or greater knowledge. Their own use of language will always be adequate for their own thinking.

However, there is no denying that linguistic style and "correct" language have an important place in communicating with others. Children may be able to say things in their own way and make themselves understood, yet their way may be neither elegant nor "standard" and some people will hold it against them. They may be able to write things with "standard" grammar and even with elegance, but with idiosyncratic spelling, and again some people will hold it against them. There is ample justification for this. Part of the reason for standardizing grammar and spelling is precisely so that we do not notice them and can give all our attention to *what* is said, rather than to distracting aspects of how it is said.

But there is a conflict for teachers here. To the extent that children are acting intelligently, they will be paying attention to the sense of what they hear and read, and not to the detail. Somehow, we must turn their attention to the detail. This would seem to imply that they have to turn off their intelligence while they do this. Indeed, that is the way "correct" grammar and spelling have most often been taught.

Teachers' attitudes to conventions like this might be characterized as "running scared"—in the sense that, since there is only one right way, explorations of other ways must be avoided at all costs. But why not encourage explorations in these matters, just as teachers encourage exploration in other areas? For one thing, running scared doesn't seem to work. If seeing or hearing something the right way often enough did work, why do children keep making mistakes? Most words that they misspell are words they have already seen dozens of times. Yet no matter how often they see words spelled correctly—and rarely do they see them spelled any other way!—the correct spelling does not seem to get imprinted.

On the other hand, think of how confusing it is. Let's take a prereader, who is learning not spelling, but his letters. He has happily learned the shape of a *C*, for instance, and draws it for himself—but backward. "No!" he's told, "That's not a *C*; a *C* is like this." An hour later, in another prereading exercise dealing with shapes, he is expected to realize that a square is still a square when it is sitting on its point looking like a diamond! How can he make sense out of all that? A backward *C* looks much more like itself than does a square sitting on its point. He is meant to be intelligent when he deals with squares, moving them around and looking at them in all sorts of ways, but he is severely restrained from being intelligent in dealing with letters.

Even in learning conventions, "right ways," why not give children the chance to be intelligent. With letters, that would seem to be as simple as encouraging them to explore their shapes, just as they explore any other shapes—"Yes, you're right, that's a *C*" (a *C* would still be a *C* even if it's lying on its back!)—while at the same time pointing out that, in writing, you draw it in one position only.

In grammar, surely the same thing can be done. As linguists have made amply clear, a sentence like "Larry never got none" represents just as much knowledge of grammar as the standard "Larry didn't get any." It's just a different grammar. It can be accepted on its own terms, while at the same time other ways of saying the same thing are explored, including "Larry got none," "Larry never got any," or "Larry didn't get none." Instead of running scared of anything but the standard form, teachers can encourage the search for all possible forms that say the same thing. And the standard can be pointed out along the way.

This does seem a bit scary. By way of reassurance, let me describe an approach to spelling that has been developed in a school in Montreal, a French-speaking school called L'Ecole Nouvelle Querbes. This approach was elaborated by Albert Morf, a psychologist of the University of Montreal, formerly of Piaget's Center in Geneva. It was developed for the classroom by first-grade teachers Helene Pothier, Denise Gaudet, and Cecile Laliberte. The approach is slightly more appropriate to French than to English, but aspects of this approach could certainly be adapted to English.

The reading program starts with writing—not handwriting,

but writing to say something. A child suggests a word he wants to be able to write. Then the class together breaks it up into component sounds. Cousin, for example (I shall use the French version of the word) is broken down into *K OO Z IN*. The teacher then presents all possible ways of spelling each of those sounds: *C* or *K*; *OU* or *OO*; *S* or Z; *EIN*, *AIN*, or *IN*. (In this respect, the method is somewhat more difficult in English. In French, the "possible ways" are more regular.) The children proceed to produce all possible ways of spelling the word. "Yes, that's one way: Any more?" The more ways they get, the better. They write them on the board, and if a child has a way that is not yet on the board, he adds it to what is there. When all possible ways have been produced, the teacher tells them which is the way that is conventionally used.

Note that instead of feeling stupid for creating an unconventional spelling, the children feel clever. And they know that whoever may be dumb, in making spelling such an arbitrary exercise, it's not they! They also know, just as well as any other child, that there is only one correct way to write any given word, and this way is underlined in their notebooks, among all the possible ways. Moreover, as time goes on they develop greater and greater ability to guess, for themselves, which is likely to be the conventional way.

At the same time, the emphasis in general is on their saying what they have to say through writing. By the time they have built up a collection of how to write all the sounds, they can write anything adequately enough for someone to be able to read what they have said. The spelling may be unusual but it is always readable, and the writing is accepted for what it says.

In this process there is, for one thing, a proper sense of values: Writing is what it is all about. The first requirement of spelling is that writing be readable afterward, and the writing of these children always is. Then, to make it easier for readers, a single conventional spelling is learned. The expressive writing of these children is remarkable and becomes better through the six years of elementary school. But this is not the main point. In other schools of various sorts children do equally remarkable writing. The point is that these children really learn to spell, withal. They learn to spell not by avoiding wrong spellings in a panic, but by actively seeking out every possible wrong spelling! When the children start reading, they notice the spellings of new words that they read. Since they realize that any number of other spellings might have done the communications job just as well, they sit up and take notice. "Gee, is that how they spell that?"

Note too, that using a dictionary to check up on a spelling is possible only to the extent that you are able to generate possible spellings in advance. You can't get anywhere with a dictionary if you don't know how to start. These children know how to start.

Finally, just as when they see a written word they know that somebody has made an active choice about how to spell it, so when they see a written text they know that somebody—some fallible person somewhere—has made an active choice about how to write it. When one child reads out loud what he has written, the other children are active listeners. Sometimes their reaction is immediate acceptance—"*Oh, c'est beau.*" But other times they make suggestions about how else the original author might have said the same thing, and he sometimes decides to say it another way. They are, in budding form, aware of the thesis of this chapter—that the words themselves aren't the substance; they are one possible way of trying to express the substance, and they needn't be taken at face value.

NOTE

1. This chapter also appears in *The Language Arts in The Elementary School: A Forum for Focus*, edited by Martha L. King, Robert Emans, and Patricia J. Cianciolo, published by the National Council of Teachers of English, Urbana, Illinois.

REFERENCES

Bereiter, C., and Engelmann, S. *Teaching disadvantaged children in the preschool*. Englewoods Cliffs, N.J.: Prentice-Hall, Inc., 1966.

Christofedes-Papert. Construction des Intersections. Geneva: unpublished study, 1965.

Furth, H. *Thinking without language: Psychological implications of deafness*. New York: Free Press, 1966.

Labov, W. *The study of nonstandard English*. Champaign, Ill.: National Council of Teachers of English, 1970.

Oleron, P. Recherches sur le developpement mental des sourdsmuets. Paris: Centre National de Recherche Scientifique, 1957.

Piaget, J. *The language and thought of the child*. New York: Harcourt Brace Jovanovich, 1926.

Piaget, J. *Judgment and reasoning in the child*. New York: Harcourt Brace Jovanovich, 1928.

Piaget, J. *The origins of intelligence in children*. New York: International Universities Press, 1952.

Piaget, J. *Play, dreams, and imitation in childhood*. New York: Norton, 1962.

Piaget, J. *The construction of reality in the child*. New York: Basic Books, 1954.

Sinclair, H. *Langage et opérations:* Sous-systemès linguistiques et operations concrètes. Paris: Dunod, 1967.

CHAPTER 7

THE DEVELOPMENT OF OPERATIONS: A THEORETICAL AND PRACTICAL MATTER

Gilbert Voyat

The way children think and how their intelligence develops are theoretical matters. Yet these processes have very practical implications. They should be decisive in the formulation of the curriculum, the selection of instructional materials, and the teacher methodology. All of these should be thoroughly integrated, although, in fact, they are often treated as if they were independent of each other.

The development of operational thought in the child should be central to all of these: the teacher's functioning theory of pedagogy (in contrast to a verbalized one, if these two should differ), the curriculum she follows, and the materials she uses. In this chapter these relationships will be examined with special attention to the development of operations.

A Matter of Practical Considerations

The problem of understanding the relationship between cognitive theories and their educational implications raises numerous questions of practical as well as of scholarly importance, since the techniques of teaching are involved. Potentially, it concerns a very large segment of the population, not only the children involved, but also the teachers and other people who are in what could be called the business of education. That education is a business, implying a market, for the sale of educational materials, and competition among entrepreneurs, can hardly be denied. It is obvious when one notes the relationship between a paradigm of thought in psychology and its socioeconomic implications. We observe, on the one hand, theorists whose work is to discover and understand such matters as the development of the intelligence, the logic of teaching, and the potential devices which can be drawn from the studies; and, on the other hand, pragmatists, who sometimes are part of publishing companies, whose business is to make the results of theoretical ideas relevant in the realm of teaching and education.

Let us suppose that in a given school each individual is expected to follow the same curriculum guide for a year. At this point two situations are possible: First, the organization of this school leaves the teacher free to choose the material he intends to use as long as it is related to the guide—diversity will be achieved within a certain frame of reference. The second case is much more frequent: The school leaves almost no choice to the teacher in terms of the materials to be used in such a way that we observe near uniformity in the teaching of the topic. This latter case represents a large exploitable market and its greater frequency probably explains the extensive competition among publishers for the school market.

The problem, however, is not so much in the commercial profits as in the implications of the practice of imposing a uniform system, one that discourages, perhaps even stifles, creative measures on the part of the teacher and student interests as a basis for planning educational experiences.

Consider the problem in regard to remedial teaching of reading. Numerous materials, programmed materials, specific books, self-programmed instructions, etc., have been devised whose goals are to help the child to learn how to read. The method of introducing these programs and making them part of the teacher's world is very interesting. The publisher's salesman pays a visit to the superintendent of the school, demonstrates the material, and convinces him that this method has brought substantial results. As a proof of his assertion, he presents statistics, i.e., the results of experimental situations where the material has been successfully tried. In other cases, providing that the prospective customer school responds highly on prestige motivation, the representative shows a list of renowned schools that have already adopted the materials.

The superintendent is impressed by the results and decides to introduce this new material to the teachers. The teachers are made familiar with this new material in workshops. Usually the workshop takes place after class hours, although, if the matter is of sufficient importance for the superintendent, he may even modify the schedule in his school for a week or so in order to introduce it. This process is sometimes referred to as continuing education and improvement of teaching skills. Finally, the teacher presents the new material to his children. The paradox is that despite the cost and effort the material will not have changed the structure of the teaching itself: The same uniformity remains since no one at any point has questioned the organization of the school itself, its implicit cognitive theory, and the teaching method.

Regrettably, this practice is the most common. The implied assumption is that new materials alone can constitute a new curriculum sufficient to achieve significant educational improvement. We shall consider other possible practices—by selecting instead a cognitive theory that best fulfills the goal of providing good tools for education, is accessible to a majority, and is relatively easy to use and put in practice.

Theory, Education, and the Problem of Actualization of Concepts

The topic of a cognitive theory directed toward education is complex because it requires understanding the relation between three factors present in school learning situations: The first one concerns the theory itself with its concepts, its ideas, its educational tools, and its own consistency. The second one deals with the practice of education, its conception, its limitations, its potential effects, its goals, and its intent. The third one is the interaction between a theory and its actualization within a school.

Different theories lead to very different kinds of educational practices. They also lead to misunderstandings as the following example shows. In a recent book Rowland and McGuire (1971) made the following comment:

> The Genevan School of Genetic Epistemology represented by Jean Piaget and some of his colleagues, such as Barbel Inhelder in Geneva and Gilbert Voyat in the United States, seems little interested in the scientific aspects of American behaviorism, and little concerned with the problems of education (p. xvi).

What Rowland and McGuire misunderstand is precisely the nature of genetic epistemology and the problem of the development of knowledge. Surely they cannot find any statement by the three psychologists cited that in any way implies a lack of interest in education or in the scientific aspects of behaviorism. Piaget was a UNESCO consultant on education and has probably more publications on education than has the average educator. Rowland and McGuire reveal what is a very common and frequent misunderstanding among behaviorists who do not see the qualitative difference that separates the child's world from the adult's intelligence. They pay tribute to the importance of statistics in the exploration of psychological phenomena. They reduce development to a set of learning devices whose major characteristic is the manipulation of external features that are supposed to be encoded by the child's mind independently from the internal structure available to him at a given time. Superficially, this type of theory is very useful for the practice of education. It leads to numerous well formulated and controlled experiments whose relationship to the practice of education is almost immediate. This immediacy is not random.

The reason for this close relationship between behaviorist theory and educational results lies not in the materials produced for school consumption but in the theory itself: Given the fact that learning so defined consists essentially in the encoding of external features, the reinforcement of known situations, and the forced extraction of knowledge by the child from external situations, one can conceive of uniformity in education. As a matter of fact the problem of education is not that simple, and prudence exhibited by developmental psychologists in the face of the complexities of human growth and learning sometimes appears to the observer immersed in behaviorism as disinterest in the problems of education.

The Problem of Operations

Let us consider, as an example, a cognitive operation such as seriation which consists in the ordering of a set of 15 sticks of increasing length. The experiment itself consists in asking the child to order the sticks from shortest to longest. The sticks are presented in such a form that the child must engage in active exploration of the material to find differences in length and to order them. This experiment has been discussed by Piaget and Inhelder in the book *The Early Growth of Logic in the Child* (1964) where three stages in the acquisition of seriation are observed. Stage 1 (around 4 or 5 years of age) is characterized by a failure in seriation. In a first substage (1a) we observe no spontaneous attempts to organize the sticks. In a second substage (1b) the child does not provide a complete seriation but constructs small juxtaposed series without an order. In these series, children can recognize some couples where the first element is smaller than the second, and sometimes correctly seriates groups of three sticks, without coordinating them in the total configuration. The child is also able to find spontaneously the smallest stick of the whole series and sometimes the longest one. The intermediary elements are then put without order and without relation to the extremes. In short, we have here a range of possible behaviors, all having a common characteristic, the inability to build a correct seriation. Yet the idea that there is a whole range of possible behaviors is important for our purpose and we shall come back to this point.

At Stage 2 (around 4 to 6 years of age) we observe success by groping. For example, the child will begin with small uncoordinated series. At this level the child succeeds in building the seriation by groping, but he is unable to make use of a system of relations that enables him to dominate trial and error and master the multiple comparisons between two or three elements. In short the child at this stage will succeed in achieving an empirical seriation through groping behavior.

Stage 3 (around 7 years of age) is characterized by operatory success. At this level the child uses a systematic method consisting of first looking for the smallest or the longest element, then looking for the next one, and so on. However, this method is called "operatory" by Piaget since it shows that a child understands that a given stick B is simultaneously bigger than the previous one A ($B > A$) and smaller than the next one ($B < C$).[1]

Our purpose is not simply to reiterate the steps of the acquisition of seriation but to analyze this development from several points of view in relation to our goal: the relation of theory to educational practice.

Views on the Operatory Development

The image of the development of this cognitive operation follows an interesting pathway. In a sense, the acquisition of seriation deals with a restriction of the range of behaviors observed, as if the logical structure involved had become unidirectional. For instance, the possible behaviors in Stage 1 are very numerous without implying a difference of cognitive stage: small series, couples, finding of the smallest stick or the longest, etc., reflect a great range of variability within a particular level of development. The apparent uniformity achieved at Stage 3 does not come simply from psychological factors, but from logical considerations as well, and changes in the internal structure of the child. At the moment of the acquisition of seriation, we observe the influence of these several factors because an operation that is an internalized, reversible action is dependent upon a whole structure. Such being the case, the mastery of seriation must provide the child with new possibilities.

Discussing the status of a structure in a recent seminar, Piaget (1970b) stressed the importance of an operation when he stated:

> Now what is the new thing that a subject can do when the structure is achieved? Think what would happen with an experiment on transitivity at the stage of empirical seriation (Stage 2) through trial and error. From a logician's point of view seriation is a well known structure that consists of a chain of asymmetrical, connex relations. . . . At the stage of empirical seriation the child does not have transitivity. He is shown two sticks, e.g., two

matches, with stick 1 slightly longer than stick 2 and asked: "Is one longer than the other?" He will immediately reply: "One is longer and the other is shorter." Now . . . the longer stick is hidden under the table and . . . the child is asked to compare the second together with a third stick. The child's reply is: "That one (the second stick) is longer." Now he is told: "Well, is it longer or shorter or equal to the match hidden under the table?" He cannot apply his reasoning to the problem; he stops short at registering what he has observed. In contrast, when the child achieves an operatory strategy he always has transitivity available, immediately and without hesitation. . . . He reasons readily according to transitivity.

When you have the two properties (asymmetry and connexity) of seriation, you have at once the third property of transitivity. As logicians have shown us, these two properties suffice to characterize the structure of seriation. This then becomes a structure closed upon itself, that is, a structure which allows one to integrate various elements within it. These two characteristics are absent at earlier stages when the structure is still in the process of formation (pp. 5 and 6).

This interesting text shows the importance of structure as well as what is behind the idea of cognitive operation. The latter never exists in isolation and is always related to some other operation in such a way that the acquisition of one particular operation makes that of the other possible. The phenomenon can be referred to as the appearance of new properties.

Piaget's text also opens the pathway for experimental investigations whose outcome is essential for education. On one hand, the problem of the internal hierarchy of cognitive operations is at stake. That transitivity is a result of the acquisition of seriation seems to indicate a type of internal organization whose complete knowledge could be of great help to teachers. We observe also that the status of cognitive development is not easy to delineate.

For instance, if one asks what the origin of transitivity is, one finds that this specific concept, crucial for the understanding of number, does not come directly from a mastery of seriation (i.e., at the stage of concrete operations). As a matter of fact, at the end of the sensori-motor period of intelligence, the child possesses what Piaget has called the "practical group of displacements," which consists essentially in a practical knowledge, or knowledge in actions. It provides a framework of practical space. The child has acquired it in order, for instance, to move about in his home; that is, he will master the space to a desired object, removing or circumventing obstacles. In the group of displacements we observe the presence of a structure whose components contain not only a direct, inverse, identical, or associative operation but also the mastery of a kind of transitivity that is practical in the sense that it is limited to the realm of actions. Yet it is a real construction of the child that allows him to understand that objects including himself can move into unique trajectories even if these are broken into successive sequences. The point of the matter is that practical transitivity exists at a sensori-motor level.

The understanding of the permanency of objects also deals with the mastery of transitivity. This level of transitivity nevertheless has obvious limitations: It is limited to the actions of the child and to the immediate situation, and it is not abstract. At the level of concrete operations, we observe a reconstruction of the concept of transitivity which is new only in its abstract meaning. We must find the reason for such a delay since practical transitivity is acquired around 2 years of age and abstract transitivity around 7 years.

The Problems of Reorganization and Pathway of Development

The first problem, that of reorganization of understandings, deals with the logic of the cognitive transformation, hence, the transition from one stage to another. The second, the direction or sequence, is that of continuity and discontinuity within development. Neither of these problems is easy to answer, but the previous case shows that there cannot be a direct one-to-one correspondence between a theory and a technique of education. When we return to seriation itself, the image that we get from its development does not consist of a simple quantitative change where the child would grasp the concept of seriation with a few sticks at the beginning and increasingly more sticks as he becomes older. On the contrary, the acquisition of seriation alone shows that its development consists in a real, qualitative change in the encoding of the structure of seriation itself; this change is a change of pattern of thought—a developmental change, not simply the extension of a given category of knowledge.

There are two indications of the reality of such a situation. The first is the behaviors themselves. It is quite true that a child who is at Stage 1 displays a whole range of possible behaviors, all of which can be put under the common denominator of "behaviors dealing with small juxtaposed series." These behaviors have to do with the underlying reality that what is accessible to the child is the dichotomy (such as *small* and *big*) or a trichotomy (such as *small*, *medium*, and *big*). During this stage the child organizes his world as a function of this category of thinking, which we shall call a "paradigm," "pattern of thought," or "creode," borrowing the word from embryology as it has been several times proposed by Piaget and Inhelder. Waddington (1960) defines the term as follows:

> I have proposed the word "creode" derived from the Greek words "necessity" and "a path" as a name for such time trajectories of progressive developmental change, which arise from the nature of the causal organization of their starting point. . . . Its mechanism does not involve anything strictly comparable to a "negative feedback" dependent on a predetermined end-point (p. 82).

Waddington's comment is important. Developmental changes can occur without a dependency on a predetermined end-point. Here we have to distinguish two points of view: The first one is that of the adult (psychologist, teacher, experimenter, or parent) who knows the meaning and possesses the components of the structure (e.g., seriation), since the adult is already at a formal level of thinking and therefore can recognize what is lacking in a child's performance. The temptation to provide the child the necessary feedback is great, but it would disregard the appropriateness and the validity of the child's own point of view, for the child has no a priori knowledge of what seriation consists of, so long as he himself has not achieved it.

In other words, the child is in the seemingly paradoxical situation of undergoing developmental changes without knowing where they are going to lead him and what they are to make possible. For instance, if a child at Stage 1 for seriation were to have his errors pointed out, he would not understand the adult's point of view.

The mistakes that the child perceives are dependent upon his own cognitive level and not upon the level projected by the adult. This fact is relevant not only in terms of the perception of errors but also in terms of the apprehension of the structure of seriation itself. To say that the world of the child is qualitatively different from the adult is not merely a statement about style or an image; it reflects a reality.

Thus, we see that behaviors provide evidence of the qualitative changes in the child necessary for operatory development. Evidence from learning situations also demonstrates that something more than a quantitative change is required as a precursor to operational thinking. For instance, in one of our experiments dealing with the cognitive development of Sioux children (Voyat, in press), we were interested in the educational as well as the theoretical implications of our findings. In order to clarify the problem of learning and development we introduced a change in the experiment of seriation. After having recorded the spontaneous behaviors of the child, that is, his own spontaneous thinking (as it is done in Geneva), we introduced an interesting modification: We tried to discover the ways in which a child could use the information provided to him. In particular, we insisted upon a checking of each particular position by the child himself, told him when there was an error, and allowed him to correct it.

We distinguished four different ways in which the child integrated information: (1) no learning exhibited; (2) local learning; (3) generalized learning; and (4) finally success, i.e., no learning necessary. Among the results were the following: The children showed almost a superimposition of their modality of learning and their particular operatory stage, that is, children were unable to learn more than what their operatory level allowed them. Most of the children who were able to demonstrate local learning were not able to enhance their performance in the direction of a higher level of operativity. In other words, even if one can provide the child with information that he can use within a given context such as seriation, he is not going to surpass his performance; rather, one must wait for the child's operatory structure to develop to observe incidence of other modalities of learning.

For instance, children at Stage 1 at most demonstrated local learning and only 30 percent of them achieved even that level of information integration. In short, our "learning experiment" was only an apparent one. Merely providing feedback is not sufficient to change the ability of a child to integrate information. However, what we found illustrates quite clearly the strong parallel that exists between an ability to integrate information and a particular level of thinking.

This example serves also to illustrate the difference that exists between learning and development. As Piaget (1964) states:

I would like to make clear the difference between two problems: The problem of development in general, and the problem of learning. I think these problems are very different, although some

> people do not make this distinction. The development of knowledge is a spontaneous process, tied to the whole process of embryogenesis. Embryogenesis concerns the development of the body, but it concerns as well the development of the nervous system and the development of mental functions: in the case of development of knowledge in children, embryogenesis ends only in adulthood. It is a total developmental process, which we must restitute in its general biological context. In other words, development is a process which concerns the totality of the structure of knowledge.
>
> Learning presents the opposite case. In general learning is provoked by situations—provoked by a psychological experimenter; or by a teacher, with respect to some didactic point; or by an internal situation. It is provoked, in general as opposed to spontaneous. In addition, it is a limited process—limited to a single problem or to a single structure (p.p. 7 and 8).

The point is that development and learning cannot be equated: Development deals with a change of paradigms of thought; learning takes place within a particular paradigm. Even with an apparently simple structure such as seriation, the developmental pattern does not consist in an enlargement of a particular category such as the dichotomy or the trichotomy which would extend itself to correspond finally to the operation of seriation itself, with its systematic method and with the added property of transitivity. On the contrary, what is observed at the moment of the acquisition of seriation is a profound change in the quality of thinking. It is not that the preoperatory child is a purely perceptual being, misled by the external visible features of a configuration and concerned with particular states of reality without relating them. In a way, the preoperatory child is limited by all these properties. Yet Piaget deals with the problem in a certain way: in seriation, for instance, he conceives of the development of understanding of seriation in terms of dichotomy. These categories of thinking are logical ones in a fundamental sense. Dichotomies such as *big* and *small* have relevance not only for seriation but also for a great number of other situations. To conceive of the world in these terms is, in fact, a useful way to deal with many situations. In the child's experience this particular way to apprehend reality is often pertinent—he sees his parents who are taller than he is, he sees small and big cars, and so on—and there should be no surprise that he will also extend dichotomy to a seriation. The fact is that dichotomy is a very general concept leading to a strong paradigm that is generalized.

Piaget has said that once the child possesses the operation of seriation he becomes more of a logician and has opened himself to a whole range of new behaviors, such as transitivity, which were inaccessible to him previously. The new paradigm has indeed produced a radical change in the apprehension of reality for the child. He is able to construct, understand, and cope with new relationships among objects that were not possible before. Indeed the acquisition of seriation is a Copernican revolution for the child. Such is the case for the acquisition of any concrete operation such as conservation, causality, speed, or time. Yet one could also argue that at a preoperational level the range of possibilities is as extensive as at the operational level. This fact is true, for the child at every level is a creative being.

Educational Considerations

What is the importance of Piaget's concept of the development of knowledge for the process of education? In terms of teaching, understanding what operations are, or what the status of the development is, leads to important considerations: The first consists of understanding that, since an operation has a very general meaning, it cannot be reduced to a particular material. It is certainly irresponsible to argue that no material should be given. The point is that seriation, for example, can be achieved through a variety of particular objects all of which will have the property of corresponding to an asymmetrical, connex configuration. Seriation is not embedded in one restricted form. On the contrary, the creativity of the teacher as well as the spontaneity of the child can be fundamentally respected. The necessary condition is that the teacher understands the underlying principle of the operations, a prerequisite that is more important than his being made familiar with a particular type of material. Operations such as conservation, seriation, or classification can lead to a whole range of materials that the teacher himself can promote.

We are dealing with a kind of Copernican revolution for the teacher as well as for the child as soon as it is understood that the essential substance is not the material but the concepts themselves. This learning should not be restricted to educators alone but should be suggested to parents. For instance, two matches, cigarettes, or pencils—any 2 objects of equal length which can be compared—can serve for the demonstration of conservation of length. Any number of objects serve to promote the differentiation between the spatial organization and the concept of sum in the realm of numbers, stones, marbles, circles, etc.

The second important consideration is related to the distinction between development and learning. The problem of understanding the internal hierarchy of cognitive development can lead to fruitful learning experiments such as Wohlwill (1959) in the realm of conservation of number. He has shown that learning is possible if it bases the more complex structures on simpler ones, that is, if one follows the natural relationship between operation and the development of structures, and does not simply impose external reinforcements on concepts the child cannot yet assimilate.

From this follows a third consideration that is equally fundamental, for it relates to the establishment of curricula. In this respect, Piaget made the following points in his book *Science of Education and the Psychology of the Child* (1970a):

The symbolic function thus enables the sensorimotor intelligence to extend itself by means of thought, but there exist, on the other hand, two circumstances that delay the formation of mental operations proper, so that during the whole of this second period intelligent thought remains preoperational.

The first of these circumstances is the time that it takes to interiorize actions as thought, since it is much more difficult to represent the unfolding of an action and its results to oneself in terms of thought than to limit oneself to a material execution of it; for example to impose a rotation on a square in thought, alone, while representing to oneself every ninety degrees the position of the variously colored sides, is quite different from turning the square physically and observing the effects. . . .

In the second place, this reconstruction presupposes a continual decentering process that is much broader in scope than on the sensorimotor level (p. 31).

If such is the case, then any curriculum should simultaneously promote conditions for the child to discover the relationship between his actions and operations as well as establish conditions for decentration.

Conclusions

Such goals are not easy to achieve. Some of the reasons lie in the organization of the schools themselves, as well as in the children's and parent's reactions to them. As Himmelweit and Swift (1969) state:

Children's reactions to school are influenced not only by the learning modes and general motivation that they bring to school (for which the primary influence will have been the family) but also by the different ideas that they have developed about the school (p. 162).

Another factor concerns the teachers themselves whose professional development is supposed to include a psychological background. Yet the idea that a teaching situation provides an outlet for a teacher's own creativity as well as for the spontaneous activity of the child is still not well developed. In fact, the teacher should be convinced that his work holds out the possibility of continuous self-reorganization, new construction, and discoveries, and that his discipline contains an infinite number of possibilities for theoretical deepening and technical improvement. What makes him a teacher in the most profound sense is not a particular material but a real understanding of the problems of intellectual development, relevance for his work, their practical implications, and the recognition that he alone can make his work a meaningful way of transmitting knowledge.

We still lack important experimental knowledge in regard to the logic of development itself. Psychologists and teachers must be united in their attempt to discover the basic laws of cognitive transformations. Such discoveries, we believe, can be achieved only through cooperation between psychologists and educators. There should be no gap between theory and practice: One is the necessary counterpart of the other. The emphases are different but the goals are alike—the understanding of the child's world. Neither theory nor education alone is sufficient to provide it. It is in their interaction that the solutions will be found.

The real problem is an epistemological one. It consists of the distinction between two possible ways in which intelligence is conceived. On the one hand, intelligence can be seen as the summation of particular elements—a set of specific factors—that are studied one by one. On the other hand, intelligence can be conceived in terms of general operations in which the fundamental idea is precisely that of structure. Sigel (1969) describes it as follows:

> In essence Piaget's conceptualization of the psychology of intelligence is developmental in format, substantive in content, and operational in behavior. These characteristics make the theory eminently germane, if not essential, for education (p. 466).

An important lesson can be learned from the development of operations. Because of their general meaning and application and because they are not restricted to particular contents, they simultaneously promote the activity and creativity of the child as well as of the teacher.

NOTE

1. The discovery of transitivity, at about 7 or 8, permits deductions like the following: $A = B$, $B = C$, *therefore* $A = C$; *or* $A < B$, $B < C$, therefore $A < C$ (Piaget, 1969, p. 143).

REFERENCES

Himmelweit, H. T., and Swift, B. A. A model for the understanding of school as a socializing agent. In P. Mussen, J. Langer, and M. Covington (eds.), *Trends and issues in developmental psychology*. New York: Holt, Rinehart and Winston, 1969.

Piaget, J. Development and learning. In R. E. Ripple and V. N. Rockcastle (eds.), *Piaget rediscovered: A report of the conference on cognitive studies and curriculum development*. Ithaca, N.Y.: Cornell University School of Education, 1964.

Piaget, J. *Psychology of intelligence*. Totowa, N.J.: Littlefield, Adams, 1969; Routledge and Kegan Paul, 1950.

Piaget, J. *Science of education and the psychology of the child*. New York: Orion Press, 1970(a); Paris: Denoel, 1969.

Piaget, J. Proceedings. Unpublished manuscript. Washington, D.C.: Catholic University of America, 1970(b).

Piaget, J., and Inhelder, B. *The early growth of logic in the child*. New York: Harper and Row, 1964.

Rowland, G. T., and McGuire, C. *The mind of man*. Englewood Cliffs, N.J.: Prentice-Hall, 1971.

Sigel, I. W. The Piagetian system and the world of education. In D. Elkind and J. Flavell (eds.), *Studies in cognitive development: Essays in honor of Jean Piaget*. New York: Oxford University Press, 1969.

Voyat, G. *The forgotten people: Study of the cognitive development of Sioux Indians*, in press.

Waddington, C. H. *The ethical animal*. London: George Allen and Unwin, 1960.

Wohlwill, J. F. Un essai d'apprentissage dans le domaine de la conservation du nombre en l'apprentissage des structures logiques. Etudes d'Epistémologie Génétique. Paris: Presses Universitaires de France, 1959. Vol. 9, pp. 125–135.

PART III

The Developing Teacher

But from the point of view of the teachers and their social situation, those old educational conceptions, having made the teachers into mere transmitters of elementary or only slightly more than elementary general knowledge, without allowing them any opportunity for initiative and even less for research and discovery, have thereby imprisoned them in their present lowly status.

(Piaget, 1970, p. 124)

EDITORS' INTRODUCTION

Piaget's monumental work has been primarily directed to epistemology, to the study of the development of knowledge. The breadth of his investigations and the brilliance of his theoretical formulations are staggering. His research has engaged him in such areas as the child's language, causal reasoning, moral judgment, imagery, mathematics, physics, probability, logic, play, dreams, and perception, to name only a few. The scope and stature of his work are now resulting in a burgeoning interest not only in the theory and experiments he has conducted, but in their possible application to education. Although he has not addressed himself to such applications, he has certainly been aware of the important relationships that might be established between what he has learned about intellectual growth and possible implications for schooling.

As early as 1935 in an essay recently reprinted (1970), Piaget reported on the advances in genetic psychology, indicating their importance as a source of new methods in the field of pedagogy. In this publication, devoted to summing up and evaluating developments in education between 1930 and 1965, he discussed the reasons for the very slow movement toward a science of education in contrast with the profound new developments in psychology and sociology. He asked, in effect, where are the counterparts in education to the new knowledge about individuals and social institutions?

In this section of the book, our aim will be to bridge the present knowledge regarding the developing mind and the developing child with some applications to classroom and the role of the teacher.

We begin with an overall analysis scheme for open-system educational programs utilizing Piagetian theory. In Chapter 8 Wickens contrasts the primary features of closed and open systems of education, the former generally associated with the *S-R* model and emphasis on traditional, hierarchically organized learning of content; the latter on the individual learner and the interactional processes among children, teachers, and the broader educational environment. He uses three basic features drawn from the work of Piaget to provide a theoretical framework for open-system programs. These include the role of active involvement in learning on the part of the student, the maintenance of a social system in the classroom as an integral part of learning, and finally, the progressive development of representation.

Kamii in Chapter 9 reiterates two of the features described by Wickens, namely (1) that learning has to be an active process, because knowledge is a construction from within; and (2) that social interactions among children in school are of prime importance for learning. She suggests that at all school levels, but particularly for the preoperational child, understanding and dealing with concrete experiences, not just vocabulary acquisition, are the bases of language. She emphasizes the need for the child's development of a cognitive framework that enables him to think rather than his need for an accumulation of isolated skills. Kamii next develops the distinction between physical knowledge and logico-mathematical knowledge and the ramifications for the child's learning including the place of language in that learning process. Her concluding description of the teacher's role provides valuable guidelines for new approaches to teaching.

In Chapter 10 Kamii elaborates on Piaget's principles of interactionism and constructivism with a discussion of his theory of intelligence, the biological origin of intelligence, and the development of structures. She describes the several influences that affect the child's own development of intellectual structures, relating these to the nature of his educational experiences. Illustrations are drawn from Kamii's extensive experience with the development of a preschool curriculum.

De Meuron describes a collaborative effort made between a psychiatric clinic in a large hospital on the lower east side of New York City and a school district administration there to provide an atmosphere of change for children and their families from an inner-city slum area. This atmosphere was developed through detailed observation and study of the affective and cognitive characteristics of the children themselves in their classrooms. Frequent meetings with a small group of kindergarten, first-grade, and second-grade teachers contributed to this effort as did medical, social, and psychological services made available to the children and families at the school site. De Meuron discusses the social and intellective functionings of the children which operate, in effect, to alienate them from all that school might offer them; then she presents specific illustrations of changes in classroom practices designed to modify the serious obstacles to learning experienced by such children.

Duckworth in Chapter 12 describes how her study of Piaget's formulations came to have relevance for work in classrooms. She points out that skills of being able to watch and listen to children yield valuable insights into how children see a problem, what intellectual difficulties they encounter, and how they come to make significant steps forward in their thinking—"the having of wonderful ideas." She concludes with an account of a primary science project conducted in Africa together with the approaches she developed for evaluating the effectiveness schools can have on the continuing development of wonderful ideas in children.

In the final chapter we consider the implications of Piaget's work for the developing teacher. Curricula in classrooms influenced by his ideas are occasions for developing the mind. Effective teachers in such circumstances will tend to be active, thoughtful, resourceful adults who rely on the resilience of their minds to activate the interest and intelligence of their students. This fact has obvious implications for principals and teacher-educators. The experiences of 4 teacher-educators—Chittenden, Furth, Ginzburg, and Parker—contain useful ways of enabling teachers to achieve higher levels of self-realization.

REFERENCE

Piaget, J. *Science of education and the psychology of the child.* New York: Orion Press, 1970; Paris: Denoel, 1969.

CHAPTER 8

PIAGETIAN THEORY AS A MODEL FOR OPEN SYSTEMS OF EDUCATION

David Wickens

During the past decade, the educational field has been marked by a proliferation of educational programs designed to be models for implementation on a nationwide scale. These programs represent a wide range of approaches to the problems of education and draw on various disciplines including education, psychology, philosophy, and sociology.

Systems of Education

For purposes of comparison, these educational programs can be described as systems. The systems approach has gained prominence in education during the past decade. Most system proposals have tended to be closed ones, and in that respect they differ from the point of view in this chapter in which we shall try to apply an open-system approach to education. Of course, our descriptions of these systems are idealizations that never correspond in detail to known reality.

Within each system features can be identified that covary within limits to maintain recognizable and consistent patterns of interaction among persons and between persons and the context of the educational environment. These features govern decisions regarding identification of curriculum content, selection of materials, physical arrangement of the classroom, and interaction patterns among persons in the program. For example, in one educational setting children might sit at desks in rows occupied by pencil and paper tasks and the teacher clearly prescribe the content of the program. In another setting children might move about the room using manipulative materials and the teacher and children share in the determination of the content of the program through expressed interests.

As the various systems are examined, they can be located on a continuum ranging from a closed system at one end of the continuum to an open system at the opposite end. Although no program meets all the specifications for either extreme, consistency exists in the relative positions between programs as models for implementation. Different models may share some similar features; however, the interaction among features within the program is the determining dynamic that gives each program its unique character. For example, although two programs may depend on the use of small group work, manipulative materials, and learning through observation and discovery, one program may organize the content around a central theme for study and the other may encourage diverse projects.

Closed-System Programs

Programs based on the *S-R* scheme and the model of the reactive organism are at one end of the continuum. Educational programs within this approach exhibit the characteristics of a closed system with a primary goal to maintain a homeostatic equilibrium within the system through the use of behavioral objectives to establish criteria for behavior. Content, objectives, and instructional strategies are predetermined.

Instructional materials are prescribed and identical for all classrooms. There is minimal interaction between the system as it functions in the classroom and the wider context of an individual's life. Interaction among persons in the classroom and between persons and materials is prescribed and controlled by a system of rewards and constraints.

One feature of closed-system programs is the structuring of education as a linear-cumulative process. This process is defined through the description of sequences of instruction that are hypothesized to describe a hierarchy of conceptual development. At any point in the hierarchy, successful achievement is predicated on the mastery of prior sequences. This criterion is similar to the definition of education that has undergirded traditional educational programs and has been delineated in numerous programs that depend on workbooks and other "canned" types of materials. For example, in arithmetic, the traditionally hypothesized hierarchy of difficulty has been: counting, addition, subtraction, multiplication, and division. After this sequence has been mastered with small numbers, i.e., units and tens, it is presented again in the same order with more complex numbers.

Efficiency in learning is emphasized as a feature of this type of system. Efficiency is gauged by the rate of movement through the hierarchy of instructional sequences. Individualization of content is limited to adjusting the pace of the program, not the content, to meet the needs of an individual. Simply, the efficient model teaches the most content in the least amount of time.

Program content is structured according to traditional disciplines, e.g., arithmetic, spelling, reading. The boundaries between content areas are maintained with no emphasis on interrelating the areas through overarching ideas. Because of the requirement that all achievement criteria must be reduced to behavioral objects that are unambiguously and immediately demonstrable, the content level is necessarily limited to factual knowledge that can be recalled (e.g., naming attributes of objects) and skills that can be observed as a step-by-step process (e.g., exchanging in the subtraction operation). The learner is seen as a receiver of predetermined products of a discipline rather than an investigator of the structure underlying a discipline.

Another important feature of this approach to education is the division of roles in the program. The roles of program developer, program implementor, and program evaluator are separate and distinct. The program developer formulates the principles underlying the program and develops the program to be implemented in the classroom. The program implementor, i.e., the teacher, serves as a conduit, relaying the content from the curriculum guide to the child. Essentially, the teacher's role is passive. Responsibility for the child's failure to learn the material rests with the teacher, not with program content. Teacher ratings are based on the performance of the children.

Criteria for program evaluation are prescribed by the behavioral objectives formulated by the program developer. The program evaluator's function is twofold: (1) to identify the points in the curriculum sequences at which the required terminal behaviors are not produced, and (2) to identify teacher behaviors as they diverge from the prescribed model.

Perhaps the most distinguishing feature of this approach to education is the emphasis on conformity to a predetermined norm. Behavioral objectives represent norms for behavior at any substantial point in the program. As a consequence of this focus, products in contrast to processes are valued.

The emphasis on norms and products causes behavior to be polarized into categories of acceptable and unacceptable. Unacceptable behavior represents an obstacle to the maintenance of equilibrium within the system. Strategies for rewarding acceptable behavior and imposing constraints for unacceptable behavior, either overtly or through implication, are included in the instructional role of the adult.

In summary, closed-system programs are differentiated by several features that are interrelated and mutually consistent. First, learning is described as a linear-cumulative process. Concepts are grouped into hierarchical sequences (e.g., first, addition operations and then subtraction operations are taught, or, first, positive statements are used to define what is included in a category and then negative statements are used to define what is not included in a category). Content is limited to traditional discipline areas. Second, efficiency in learning is emphasized. Content and instructional strategies are prescribed by what is directly and immediately demonstrable. Third, roles are clearly differentiated in terms of program developer, implementor, and evaluator. The program developer defines the goals, content, and instructional strategies. The implementor plays a passive role as conveyor of the prescribed program to the learner. The evaluator determines when teacher and child behaviors diverge from the program expectations. Fourth, behavioral objectives are established that define norms for achievement at any point in the program. Fifth, behavior is categorized as acceptable and unacceptable and a system of rewards and constraints is prescribed as part of the instructional strategies used by the adult.

Open-System Programs

When the opposite end of the continuum for comparing systems is approached, basic notions about the individual and educational contexts change. The resulting educational programs are based on the model of man as an active personality system. Individual differences are emphasized as well as are aspects of human functioning that would be considered non-utilitarian in the closed system.

An individual functioning in an open-system program is recognized as an active personality system engaged constantly in active transactions with sources both within and outside of the educational setting. Maintenance of an equilibrium in terms of homogeneous expectations for behavior and achievement is not appropriate for this type of system, which also seeks to accommodate to the diversity in human functioning as a primary goal. Contingencies that are excluded from consideration in the closed system are of primary importance in the open system. For example, the expressed interests of individuals, the cultural and ethnic characteristics of a community, and the ecology of the geographic setting of a community are variables among those that are considered in the development of a program. Continuous and unprescribed interplay among persons in the program, between persons and the educational context, and between the educational context in the classroom and outside sources of influence, results in a continuous state of disequilibrium that is identified as a steady state. It is characterized by continual reorganization of the curriculum and reconstruction of the environment as needs are assessed.

Another important shift occurs with movement along the continuum from a closed system to an open system. The primary focus on product in the closed system is replaced by a primary focus on process in the open system. With the educational process as the major focus for consideration, evaluation techniques must emphasize the network of interrelationships among persons and the use of the educational environment by individuals. The general antipathy to summative evaluation techniques by advocates of open-system educational programs is rooted in the emphasis on process. Not only the limited focus of summative evaluation techniques but also the instruments that are so often alien to the experience and demonstration of knowledge by the student in the classroom situation are evidence of the inappropriateness of summative evaluation for an open system.

A qualitatively different conception of the learning process underlies the emphasis on process in the open system. The open-system program is based on a model that hypothesizes the development of increasingly complex cognitive structures through continual reorganization. As new experiences are assimilated, cognitive structures are reorganized and an individual's perceptions are altered. This reciprocal process depends on a continually widening repertoire of experience at each stage of growth. The vertical progression through stages of development is complemented by a horizontal development exploiting the potential for deeper development at each level. The emphasis on acceleration through efficiency in learning is replaced by an emphasis on the application of knowledge to a wider range of situations in service of an individual's interests.

The emphasis on process has important ramifications for approaching the study of a discipline. Within an open-system program, the individual is not a receiver of the products of a discipline. Instead, he is concerned with developing an understanding of the structures that underlie all forms of knowledge. The interrelationship of disciplines is achieved through an understanding of the structural characteristics of knowledge. Barriers between the traditional discipline areas are broken down and disciplines are interrelated through the use of work situations that require the synthesis of knowledge from a variety of disciplines. This approach represents a reversal of traditional educational philosophy in which only facts and skills are taught to all except those who reach the upper educational levels and are accorded the privilege to hypothesize and speculate about the structural characteristics of a discipline.

Because a wider range of human functioning is considered in an open system, norms for achievement serve as dubious guidelines, and expectations for finished products and for terminal behaviors are less specific. Observations of the process are important as material with which to diagnose a child's level of understanding and ability to function. For example, inability to exchange[1] in the subtraction operation might be an indication of a lack of understanding of the basic grouping patterns underlying a decimal system. Experience with grouping objects and reorganizing groupings would be more relevant than experience with the rote subtraction operation. Similarly, not only would inability to recognize words be remediated by decoding practice, but more practice would be arranged in using words to identify block constructions or their functions, recognizing signs with specific meaning relevant to the learner's daily experience, or associating words with other representations of experience produced by the child.

Because the open system is characterized by continual reorganization in response to the needs of the learner, the teacher must assume the roles of program developer and implementor. Additionally, because the teacher must exercise judgment about the needs of the individuals within the constantly changing environment, he also assumes some aspects of the evaluator's role. For example, the teacher is expected to notice when a child is ready to begin to learn about reading based on an assessment of skills for discrimination of auditory and perceptual clues. At the same time, the teacher is expected to assess the child's motivation to read. He also must be aware of the characteristics of the child's learning style and ways of organizing knowledge to determine which approach to reading is most effective. With this information, the teacher matches materials and experiences to promote learning. Finally, the teacher continually assesses the child's progress. In the event the child does not show progress, the teacher reassesses his previous decisions.

In summary, open-system educational programs are distinguished by the following features: First, there is a holistic conception of the learner and the educational environment with export to and import from sources outside of the classroom. The process the learner experiences in the educational environment should be usable outside the educational environment. Second, an open system is characterized by a steady state of disequilibrium in response to changing interests and needs of individuals in the program. This is a state necessary to the dynamics of a successful program. Third, the focus for evaluation of the program is the network of interrelationships among individuals in the program and between individuals and the educational context. The focus on process deemphasizes products. Fourth, the roles of program developer, implementor, and evaluator are combined within the role of the teacher.

The Relationship of an Open-System Program to Piagetian Theory

Obviously, the most formidable problem in describing the processes that characterize an open-system educational program is first to identify features that are central to the dynamics underlying the process. These features pertain to both the construction of the educational environment and the behavior of individuals in the program. They should covary within limits to produce recognizable patterns of behavior that consistently characterize open-system programs.

The Piagetian theory of development provides one useful source for the identification of features. Although his is an epistemological theory, it is possible to identify some basic ideas that pervade the theory and seem to be particularly relevant to the task of establishing a framework for the identification of features common to all open-system programs. The following ideas provide that framework:

1. Piaget conceives of action as a pervasive mode for development including forms of internalized action at later stages of development.
2. Piaget emphasizes the primary role of the socialization process in organization of knowledge and communication if it is to be effective.
3. Piaget describes a continuum of development characterized simultaneously by less dependence on spatial and temporal contiguity and an increasing dissociation between general form and particular content. In other words, the intrinsic relationship between the real object and knowledge about it is diminished in favor of the development of abstract forms for organization of knowledge.

LEARNING THROUGH ACTIVE INVOLVEMENT

Piaget has redefined intelligence as an adaptive process learned through active transactions with persons and features of the environment. The characteristics of involvement change with the developmental process. Younger children use practical schemes to coordinate external actions and learn through the use of manipulative materials and direct experience. Older children use operational schemes to coordinate generalizations about experience. As language assumes a more central role in the learning process, the child is able to use vicarious experiencing as a vehicle for learning. Throughout the developmental process, however, active involvement is essential for the development of cognitive strategies through which experience can be organized.

The major problem in analyzing the level of active involvement in an open-system program is to define the characteristics of opportunities for active involvement which are presented to the learner. Open systems maintain a steady state of disequilibrium through exchange with the outside and continual reorganization of the curriculum and reconstruction of the environment.

The role of the teacher. A primary source for understanding the characteristics of active involvement can be found in an analysis of the constraints and options that a teacher offers the learner. In a closed-system program, the teacher's role is to convey effectively the prescribed content to the learner. The learner is expected to conform to the expectations for interest and achievement that the teacher establishes. Within this framework, identification with the learning process depends on identification with the adult who establishes the criteria for successful performance. The learner is given no options in the initial selection of content or in the subsequent options to attend. Constraints available to the teacher consist of a system for administering rewards and punishments to reinforce his central role.

In an open-system program, the teacher's function is to create an environment in which the learner is interested in exploring and studying about what is relevant to his interests. The teacher helps the child to identify with the learning process as a means for serving his own interests and needs. The adult is no longer an intermediary through whom learning is prescribed and performance is accredited. The teacher has choices available for offering options and imposing constraints on the learner. One choice concerns the level at which the constraint or option is used. For example, a teacher might impose a constraint at the first level of determining the course of study for a group (e.g., a primitive culture). However, at subsequent levels, options can be offered about what aspects of a particular primitive culture each person chose to study. Conversely, options can be offered at the first level and constraints about procedure can be imposed at subsequent levels.

Another choice deals with the extension of options and constraints. The teacher has the choice of applying constraints at a particular level to one person while simultaneously offering options to another person. The teacher's decision is dependent on a diagnosis of what is most helpful to the individual at the moment.

The difference between the teacher's role in closed- and

open-system programs in terms of the use of options and constraints has important ramifications for the difference in the type of control function that the teacher serves. If the teacher is the prescriber of standards and content, this control and organization of the classroom must be centralized in the person of the teacher. On the other hand, if the teacher is concerned with matching content and interests, the control and organization of the classroom are vested in the student's motivation to learn.

The nature of the curriculum. An analysis of curriculum content to include consideration of several factors is another source of information about the characteristics of opportunities for active involvement that are offered within a program.

First, an analysis must consider the appropriateness of content to a particular developmental level in terms of the distance between experience and learning, i.e., the need to have direct experience versus the ability to use vicarious experience for learning.

Second, the relation of the classroom environment to the wider context of the learner's life must be considered, in other words, the inside-outside relationship of the open-system classroom with the outside world. Active involvement in the learning process depends on the child's interest in exploring the learning environment. Consequently, consideration of the child's interests and needs becomes a central issue for curriculum development in an open system. Similarly, if the curriculum is to be considered relevant for and by the learner and if the educational program is intended to extend to other aspects of the learner's life outside of the classroom, use of the larger environment of the learner becomes an important consideration for program developers and implementors.

Third, in addition to the range of choices offered for the pursuit of interests by an individual, the degree to which the various interest areas offer the opportunity to integrate experiences must also be considered. In other words, to what extent can an interest be developed and integrated through the options available in a variety of interest areas? For example, if a group is studying a primitive culture, the science, dramatic play, construction, and reading areas should offer opportunities to investigate various aspects of the central theme.

The environmental context of an open-system program also offers certain features for the analysis of opportunities for active involvement. A primary consideration is the number of interest areas in a classroom as an indication of the options an individual has for pursuit of interests. Another is the adequacy of materials within a particular area for the development of interests. Adequacy includes the variety of materials for developing inductive thought processes and, more importantly, the extent to which the materials are nonrepresentative and able to be combined. Nonrepresentative materials offer an opportunity for the learner to project meaning onto a material and to organize experience through the use of the material. Combination of a variety of materials allows the possibility for increasing the complexity of cognitive organization. For example, in a setting for young children, planks can be used in combination with climbing bars in order to extend play. Similarly, sand can be combined with measuring containers and water for elaboration of play patterns. In a setting for older children, scales, measuring containers, floating and sinking materials, and water can be combined for a variety of experiments to investigate the relative properties of objects and materials.

In summary, the opportunities for active involvement in a program can be analyzed in terms of the pattern of options and constraints that the curriculum and the environmental context offer to an individual. Options are represented by opportunities that open a situation to exploration and expression. Constraints are represented by situations in which content and response are prescribed by the adult.

THE SOCIALIZATION PROCESS

Socialization and the maintenance of the social system in the classroom, as a fundamental process for learning, are concerned with the means by which the teacher helps the child make choices and self-regulate his behavior in relationship to his peers. The developmental process leading from a morality of constraint to a morality based on mutuality is implemented in the classroom as the process by which teacher-defined rules are replaced by child-defined rules generated by mutual respect and awareness of the reciprocal relations underlying cooperative group functioning.

Two major techniques derived from Piagetian theory are relevant to the socialization process and serve as a framework for analyzing the process in operational terms. The first concerns reinforcement; the second concerns developing the child's ability to be effective in relationships based on mutuality through the use of logic.

The open system is premised on learning through active participation with emphasis on the autonomy and freedom of the child. Guidance by the teacher is a crucial factor in developing this atmosphere for learning. It is a means for involving the child in the diagnostic process and, subsequently, in the goal-setting process. This is enacted by behavior on the part of the teacher which is organized to elicit self-evaluation from the child. For example, if a child asks a teacher to look at a piece of completed work, the teacher avoids statements implying a value judgment. Instead, the teacher asks the child to review the process through a question such as: "How did you do that?" or, "Can you tell me something about this?" If the child asks the teacher for a value judgment, the teacher replies by asking the child about his feelings, such as, "How do you feel about this?" Moreover, responses are respected and the teacher does not attempt to discredit the child's assessment. If the assessment is negative or positive, the teacher attempts to relate it to planning a future experience.

Other forms of guidance include feedback to the child that behavior is inappropriate. The child's perception of a situation is elicited and criteria for behavior are examined with the child by the teacher's logically analyzing the situation and developing reasonable expectations for everyone in the maintenance of a viable social system. Social reinforcement is implied in the teacher's interest and concern for the individual.

The use of logical reasoning for mediating teacher-child and child-child interaction is another important feature of socialization in the open-system program. It evolves as a function of establishing Piagetian notions of identity and conservation in the classroom operation. In this context, the term "logic" applies to a wide spectrum of mental activities including anticipating the effects of certain possible behaviors of the child.

Identity as a principle of logic requires that the definition of terms remains constant and relations of logic within a situation can be established independently of the participants. Conservation is the equivalent of identity in the realm of physical experience. Both terms refer to the discrimination of constants and variables in a transaction or transformation process. They underlie development of the ability to use hindsight in order to anticipate probabilities in the future.

Identification of identity and conservation elements is facilitated by a combination of maturation and experience. The experiential basis for discrimination of these elements is the substance of the school curriculum. Using the repertoire of experience provided by a curriculum, the teacher helps the children continually to identify recurrences and regularities as identity and conservation elements. As development proceeds, this basic repertoire of experience is generalized through language to enable vicarious experience of a more expanded and differentiated repertoire. Language facility develops and is controlled increasingly by logical necessity. Behavior, in turn, is controlled increasingly by language.

The use of logic in socialization requires an orientation by the teacher to posing problems that allow for inductive thinking. It emphasizes hypothesis-making with the special purpose of drawing attention to causal relationships. A notion of the causal relationships that underlie the probability of the occurrence of an event is the basis for predictability in social systems. This process is repeatedly rehearsed in every area of the curriculum whenever hypothesis formulation is introduced to accompany other experiences. For example, "If you mix blue and red, then what color will you have?" With older groups, the problem can be posed in terms of ecology. "If the water supply in an arid area could be increased through irrigation, what is the range of effects that would be produced?"

Social functioning within a peer group, in particular, is developed through constant reiteration of the causal relationships of behavior. For example, "If you hit Johnny, then what do you think he will do?" "If you want a turn on the swing, then what do you do?"

The development of a commitment to a logical necessity in the social context implies a teacher-child relationship based on mutuality. The process of interaction is a continual reaffirmation by both adult and child of the use of logic for social interaction. The structure of logic varies according to developmental levels. For example, at the 3-year-old level, if one child has all of the fire engines and another child wants one, the teacher recognizes sharing of quantities as the understandable logic underlying the situation. However, if a child at the 5-year-old level has constructed a firehouse and claimed all the fire engines, the logic of the situation justifies his possession. If another child wants a fire engine, he will have to construct another firehouse and enter into the logic of the situation.

Approaching the 7-year-old level, participation in rule-bound situations such as games and responsibilities for the maintenance of the classroom become an important experience for the child's understanding the logic underlying social communication.

In each case the logic of the classroom situation parallels the logic of social functioning in the larger world. Socialization, as a process, is based on the notion that school is life itself rather than a preparation for life. The maintenance of a social system is dependent on each person's taking an active role in living within that system.

USES OF REPRESENTATION

Piaget has described a development from egocentric and personal forms of representation to more socialized forms with universal meaning. This is a many faceted development that includes stages in which physical, pictorial, and formal symbolic representations are predominant. It is also characterized by a movement from dependence on spatiotemporal contiguity and redundancy of information to independence of space and time in thought processes and the ability to perceive general form in specific examples.

Two important implications for the analysis of programs emerge within this framework for development. First, the direction of development of forms for communication is from the personal to the universal, from the egocentric to the socialized. Representational thought begins with the use of private, nonverbal forms that are internalized through the use of imitation and the use of physical objects as signifiers. Gradually, the use of pictorial representation is assimilated to the repertoire of forms, and finally formal symbolic forms are also included. The important point is that representational thought does not begin with the use of abstract, formal representational forms.

Second, as the process of representation proceeds and more sophisticated forms are included in the individual's repertoire of experience, all forms of representation are mapped together in a mutually assisting relationship. Ideally, experience in one form of representation is reinforced by analogous experiences through other forms. An individual's organization of knowledge should encompass physical, pictorial, and formal symbolic forms of experience. The range of experiences offered within the program in terms of the variety of forms available for representing and organizing knowledge is a primary consideration for analysis. The variety of materials available for expression and the process by which materials are combined for expression of a single experience through different modes for representation are also important aspects of representational activity.

Examples of the former include materials that can be used for construction activities such as building blocks, multibase mathematics blocks, or models of physical systems. Examples also include materials available for pictorial representation such as paints, photographs, and maps. At the level of formal symbolic materials, books, records, and charts are examples.

Possibilities for use of materials to establish correspondence between forms of representation of experience vary with age. In the case of younger children, a study of the community can be represented through block constructions. The same experiences can also be represented through paintings. Signs attached to the block constructions to indicate their functions are an example of mapping two modes together. Subsequent discussion of the functions and interrelationships of people and constructions has the effect of retrospective analysis and lends to the subsequent process of further clarification of ideas.

For older children, the order of modes from physical through symbolic can be reversed. For example, a study about dwellings can begin with reading about dwellings in different parts of the world. Models of the dwellings are constructed and a subsequent discussion would deal with the reasons for variations in construction and the use of materials.

Generally, as each successive form for representation of experience is employed, the effect should be reflectively to clarify the former representation and to lend more universal significance to subsequent development of the latter representation.

Concluding Comments

Open-system educational programs present unique problems for studying the specific effects that a program has on the development of an individual. Educational products are emphasized in closed-system programs and are available through prescribed behavioral objects as criteria for successful achievement. However, process is emphasized in open-system programs, and product variables are neither prescribed nor standardized for all programs within a particular model. The emphasis on process presents an especially formidable problem for evaluation. Process emphasis implies valuing the ways in which an individual interacts with his peers and adults in cooperating for mutual benefaction through exchanging ideas, utilizing each person's strengths, and respecting the feelings of each other. Process emphasis additionally implies optimal use of skills and knowledge by the learner for the promotion of individual interests within the humanistic framework of nonpredatory relationships with other human beings.

As a first step to understanding the potential range of effects a program can have on individuals, a typology of the teaching-learning process must be developed. The purpose of identifying features of an open-system educational program for analysis is an attempt to describe aspects of this type of program which are essential to the optimal functioning of a program at the operational level. These features should describe a range for interactions among persons in the program setting, guidelines for construction of the educational environment, and a range for the type of transactions that individuals are engaged in with that environment. Within this framework for observing and recording the operation of an open-system program, it seems possible to describe a program in terms of proximity to the ideal model for an open-system educational program.

NOTE

1. Exchange: i.e., in subtractions involving more than one digit in which at least one digit in the subtrahend is larger than its counterpart, requiring an exchange, e.g., in 54 — 38 the 4 is transformed to 14, in an exchange with the subtrahend, which gains 10, becoming 48.

CHAPTER 9

PEDAGOGICAL PRINCIPLES DERIVED FROM PIAGET'S THEORY: RELEVANCE FOR EDUCATIONAL PRACTICE

Constance Kamii

In this chapter[1] Piagetian pedagogical principles will be compared with theories on which much of current practice in schools appears to be based. The discussion will begin with statements of three main principles together with an elaboration of these statements expressed by Piaget about education (1935). It will continue with a focus on several of the current beliefs and practices in education that conflict with Piaget's theory. References to such current thinking will be to the general state of the art, not to the exceptions. A concluding section will consider the role of the teacher.

Basic Pedagogical Principles of Piaget

A first principle drawn from Piaget's theory is the view that learning has to be an active process, because knowledge is a construction from within. Almy et al. (1966), Chittenden (1969), Ginsburg and Opper (1969), and others have all emphasized this point. Duckworth (1964) selected the highlights of the statement Piaget made on education in 1964:

> As far as education is concerned, the chief outcome of this theory of intellectual development is a plea that children be allowed to do

> their own learning . . . You cannot further understanding in a child simply by talking to him. Good pedagogy must involve presenting the child with situations in which he himself experiments, in the broadest sense of the term—trying things out to see what happens, manipulating symbols, posing questions and seeking his own answers, reconciling what he finds one time with what he finds at another, comparing his findings with those of other children . . . (p. 2).

This statement expresses the major difference between Piaget's theory and the other theories on which current practice in education is based. Teaching is still, by and large, considered a matter of presenting the material to be learned and reinforcing the correct answers that the learner gives back to the teacher. Even when a discovery method is advocated, "discovery" usually means to discover only what the teacher wants to have discovered.

A second principle suggests the importance of social interactions among children in school. Piaget believed strongly that for intellectual development the cooperation among children is as important as the child's cooperation with adults. Without the opportunity to see the relativity of perspectives, the child remains prisoner of his own naturally egocentric point of view. A clash of convictions among children can readily cause an awareness of different points of view. Other children at similar cognitive levels can often help the child more than the adult can to move out of his egocentricity.

Teachers and teacher trainers often advocate committee work and discussions among pupils. However, current theories do not have a clear rationale for deliberately setting up the classroom to encourage children of similar cognitive ability to exchange their views. In practice, social interactions among children are *allowed* more than they are expressly *encouraged* as a means of actively involving children in juxtaposing different points of view.

A third principle points to the priority of intellectual activity based on actual experiences rather than on language. Almy et al. (1966), Duckworth (1964), and Furth (1970), for example, have pointed out that language is important, but not at the expense of thinking. Sinclair and the author (Sinclair and Kamii, 1970) have particularly insisted on the necessity of letting the preoperational child go through one stage after another of giving the "wrong" answers before expecting him to have adult logic and adult language. Sinclair has shown (Chapters 2 and 3) that the child reasons from different systems (e.g., number and space), each of which is correct in itself. We shall return to this point later in the discussion.

Teachers and teacher trainers have made some progress in recognizing the importance of concrete experiences prior to using words, but the emphasis is still on words and the correct answer that the teacher wants. At the preschool level, there is an incredible preoccupation with the teaching of language without coming to grips with how the preoperational child really thinks. In junior high school, youngsters still answer the questions at the end of the chapter by copying the book day after day. A typical question is "Define photosynthesis."

Furth (1970) recently went even further than objecting to educators' overemphasis on empty words. He took a bold and correct stand in saying that the first goal of education must be to teach thinking. He argued that reading is indeed an important tool for learning, but it has unfortunately become the preoccupation of educators to the exclusion of high-operative thinking. Schools of education have made a cliché out of individual differences, but in practice educators still continue to expect all children to learn to read in first grade, whether or not they are interested and/or "ready" for reading.

In order to develop the first principle further, we shall turn to what Piaget himself said about active learning in *Encyclopédie Française* (Piaget, 1935). This early publication is tangible evidence of how incredibly long it takes for research findings to be applied in education, even when the researcher himself sits down to write the pedagogical implications of his research. Much of what Piaget criticized then as being obsolete is still being practiced in our schools and advocated in teacher-training institutions.

With regard to active learning, Piaget pointed to Pestalozzi, Froebel, Montessori, Susan Isaacs, and others who had developed "active" methods in the sense of involving actions on things rather than listening to the teacher or looking at books. However, Piaget found various problems with each one of these "active" methods. Many of the objections he raised are still in the vanguard of educational criticism today. For example, Montessori wanted to give to children the materials that would exercise the free, spontaneous activities of the child to develop his intelligence. However, according to Piaget, her materials are based on a psychology that is not only an adult psychology but also an artificial psychology. Making cylinders fit into holes, seriating colors of various shades, discriminating the sounds made with stuffed boxes, etc., may be activities, but they are activities that take place in a limited and artificially limiting environment. Piaget suggested that in trying to free the child from clumsy adult interference, Montessori's work emphasized sensory training rather than the development of intelligence.

What Piaget said about Susan Isaacs' belief in experience is particularly relevant to current issues in early childhood education. Isaacs' school had the greatest collection of possible materials at the children's disposal, and in order to let the children organize their own experiences the teachers abstained from intervening in the children's play. Piaget wrote that in emphasizing the role of experience as the foundation for intellectual development, Isaacs overlooked the importance of the structuring, elaborating, and reasoning processes. The children learned to observe and to reason by playing freely, said Piaget, but nevertheless *some* systematization by the adult would have been helpful. There is, after all, the need for a rational, deductive activity to make sense even in scientific experiments.

In a nursery school for privileged, bright children, it is probably sufficient just to let the children have a lot of "experience," because they themselves will give structure to their experience. In a program for the disadvantaged, however, the children need structuring appropriate for them, and this requires the teacher to be in genuine contact with the child. Experience alone, no matter how rich or varied, is not sufficient unless the level of the child's development is considered.

Structuring, too, can be excessive. Piaget credited Pestalozzi, a disciple of Rousseau's, with being right in having a school that was a true society, with a sense of cooperation and responsibility among pupils who actively helped each other in their research. However, Piaget criticized Pestalozzi for being preoccupied with the idea of having to proceed from the simple to the complex in all domains of teaching. Therefore, in spite of Pestalozzi's general belief in the importance of the learner's interests and activity, his methods were characterized by schedules, the classification of the content to be taught, exercises for mental gymnastics, and a mania for demonstrations. For Piaget, young children's logic deals sometimes with undifferentiated, global wholes and sometimes with isolated parts. Therefore, we cannot always proceed from the analytic parts that to adults seem simpler than the whole.

To summarize what Piaget said about active methods, he pointed out that the criterion of what makes an "active" method active is not the external actions of the learner. He said, for example, that Socrates used an active method with language and that the characteristic of the Socratic method was to engage the learner in actively constructing his own knowledge. The task of the teacher is to figure out what the learner already knows and how he reasons in order to ask the right question at the right time so that the learner can build his own knowledge.

Piaget is usually believed to have interest only in the child's cognitive development. In 1935, however, he wrote that the goal of education is to adapt the child to the social environment of the adult. In other words, for Piaget, the purpose of education, in a broad sense, is to transform the psychobiological constitution of the child to function in a society that stresses certain social, intellectual, and moral values. The important difference is that, when Piaget talks about the social, intellectual, and moral values of the adult world, he insists that part of society's goals should be not only the transmission of old knowledge and values, but also the creation of new knowledge and values. This point is substantiated by what Piaget said about Rousseau's pedagogy.

Piaget praised Rousseau for believing that the pupil has to reinvent science rather than merely to follow its findings. However, he criticized Rousseau's pedagogy for dissociating the individual from his social milieu. In the traditional school, said Piaget, adults are the source of all morality and all truths. The child merely obeys the adult in the moral realm and recites things in the intellectual realm. In Piaget's opinion, schools must emphasize not obedience but the development of autonomy and cooperation. He criticized Rousseau for believing that a child could learn to be moral without practicing moral judgment and moral behavior with other children in school. Just as intelligence can develop only by being actively used, moral behavior, too, can develop only by being actively used in daily living. In today's modern school, "citizenship" is still all too frequently equated with "obedience."

Current Educational Practices

Turning to an examination of current education practices in the light of Piaget's theory, we see two clusters of beliefs and practices: the school's focus on teaching *skills*, and the belief that children move from *concrete experiences* to *abstract thinking*.

THE TEACHING OF SKILLS

Education has progressed at least to the point of our becoming aware of the futility of teaching encyclopedic facts. In turning away from mere facts to a focus on cognitive processes, however, educators became preoccupied with the teaching of skills. Skills traditionally have implied motor or mechanical skills, or isolated proficiencies such as typing, penmanship, or swimming. In the modern version of the report card teachers are asked to check whether or not the child has *skills*—arithmetic, visual, auditory, word attack, information-processing, and many others. There may be some justification for delineating these behaviors as skills, but the inclusion of so-called comprehension skills in reading raises questions about how the child understands the written word and whether that process can validly be defined as a skill.

Piaget's biological theory of intelligence states that intelligence is a general coherent framework within which all the parts function. It becomes clear, then, that addition skills, for example, are not mere skills. Although specific training in addition may lead children to rote learning, it is unlikely to contribute to the development of a general cognitive framework in which addition and subtraction processes can be understood, either by themselves or in relationship to each other.

In this connection S. Lai, one of our teachers who taught in our preschool for two years before teaching a class of first graders referred to as *slow*, reported the following: After teaching these children to read according to an experiential approach, she came to the conclusion that, although the mechanical skills of reading can be taught relatively easily, comprehension of what is read cannot be taught directly. These children, for instance, made little out of a sequence of pictures in the book they were supposed to be reading. They lacked the physical, social, and logico-mathematical framework required for such an understanding.

Furth (1970) insists that meaning in words and pictures is derived not from their figurative aspect but from operative knowing. Education is still preoccupied with the figurative aspects of knowledge. Reading cannot be taught just by teaching the mechanical skills any more than children can be expected to identify with the pictures and become better readers just because the white faces in *Dick and Jane* have been changed to black faces.

Even perceptual discrimination or perceptual skills require more intelligence than educators usually assume. Current practices suffer from not having made the distinction that Piaget made between perception and perceptual activities. He (Piaget, 1969; Piaget and Inhelder, 1967) emphasized the role of actions in perception. He showed, for example, that recognition of geometric shapes cannot be achieved only by perception, i.e., only by focusing on the immediately given sensory data. This recognition requires perceptual activities, i.e., the moving of the eyes, hands, or feet to center on various points of the particular spatial configuration to construct it actively. What is important in perceptual activities is not the action of moving the eyes or fingers but the active mental construction of the spatial configuration.

Let us assume that the objective to get the child to recognize a square through class actions is an appropriate one. The teacher who is aware of the difference between perception and perceptual activities probably uses haptic perception (the sense of touch) to achieve her goal. This method of teaching is not just the use of many sensory modalities for perception. Visual perception of simple shapes is rather automatic, but haptic exploration presents the child with the need to construct a spatial structure when his visual perception is not permitted, i.e., haptic perception provokes perceptual activity. Sharpening the teacher's theoretical understanding can thus result in very different teaching methods.

The comments suggest that a psychology that focuses on sensory information and motor behavior results in the conceptualization of goals and teaching methods involving skills rather than on the child's entire cognitive framework. In the next section another set of examples will be presented from a different angle to suggest again that a lack of theoretical precision results in the kind of teaching that does not maximize children's chances of developing their intelligence.

PROCEEDING FROM THE "CONCRETE" TO THE "ABSTRACT"

Educators have progressed to the point of realizing that children do not learn by simply being told or having things explained to them verbally. We now recognize the necessity of having concrete experiences first and buying lots of manipulatives from toy companies.

Few people have asked, however, what concrete experiences are, and what abstraction means. Concrete experience usually refers to any direct contact with real objects and events, and abstract thinking generally refers to the use of representation and the so-called higher order concepts. Piaget's conceptualization of experience and abstraction gives guidelines to the teacher as to how to make moment-to-moment decisions in the classroom to develop children's intelligence.

Closely related to the importance of "concrete experience" is the importance attached to "discovery." It is often said that the child learns more through direct experience, and that he learns even more if this experience is discovered rather than being offered. The pedagogical question becomes "Should we design a curriculum in which everything is taught by the discovery method?" Piaget makes a distinction between "discovery" and "invention." His favorite example of "discovery" is Columbus' discovery of America. Columbus did not invent America, he points out. America existed before Columbus discovered it. The airplane, on the other hand, was not discovered. It was invented because it did not exist before its invention.

Corresponding to the distinction between discovery and invention is the distinction between how physical knowledge and how logico-mathematical knowledge are constructed. Physical knowledge can be built by discovery, but logico-mathematical knowledge cannot. It can be built only by the child's own invention. For example, by putting a needle in water, the child can discover whether or not the needle floats. By acting on objects, he can thus discover their properties. In logico-mathematical knowledge, on the other hand, the child cannot discover from the objects themselves whether there are more brown beads or more beads in a collection. All logico-mathematical structures have to be invented, or created, by the child's own cognitive activity rather than discovered from the reaction of objects.[2]

An example from our own project will illustrate the importance of the distinction between discovery and invention. One of the many mistakes made during the first year of our project was to concentrate on logico-mathematical knowledge and to overlook physical knowledge. Another thing we did wrongly was to make up many classification kits and select the objects according to how we thought children should sort them easily. For example, we made kits for sorting by color or use (e.g., a red comb, red hairbrush, red clothespin, and red pencil; a pencil, pen, and crayon). Naturally, when the children sorted in ways for which the kit was not intended, the teachers said to them, "That's very nice, BUT . . . can you think of another way?" (i.e., the teacher's way).

In March of the first year half a dozen Piagetian friends arrived to help in the development of our curriculum. They pointed out to us that in logico-mathematical knowledge, when we contradict the child, we only make him unsure of himself, because without the necessary cognitive structure the child has no way of understanding why his way is not right. Particularly in classification, whatever criterion the child selects is right, provided he uses it consistently.

In physical knowledge, in contrast, the child is not made to feel unsure of himself if the object flatly contradicts him. If the child believes that a block will sink when it is put in water and the object contradicts him, he can understand the object's reaction as being part of its nature. We went on to speculate that an emphasis on the teaching of physical knowledge may even give a sense of security and confidence to the children by enabling them to control objects and to predict the regularity of their reactions. We knew that our children lacked initiative and curiosity and speculated that a lot of activities in physical knowledge might increase their curiosity because the objects' reactions were in themselves interesting to all young children.

With this wisdom, we ran our second year with a new group of 4 year olds, and the contrast between the first and second year was dramatic. Our visitors' most frequent comment during the second year was that our children were independent, well organized, active, and full of initiative. We thus found that the theoretical distinction between "discovery" and "invention" made a difference in the way we taught and in the "concrete experiences" children had with objects. When our teachers learned when to let the child discover and how to facilitate "invention" indirectly, our children became more confident about their ability to figure things out. They stopped looking at the teacher's face for approval and relied more on their own ability to think. The inseparable relationship between the child's cognitive development and ego development thus became evident.

Language and the importance of encouraging children to say exactly what they think assume an important place here. First, unless children tell us how they think, we cannot get the diagnostic insights that are essential for diagnostic teaching. Second, children must have sufficient confidence in their own way of experiencing things and their own process of reasoning rather than "learning" through social conformity only. For example, if the child thinks that 8 objects spread out in a top row are "more" than 8 objects clustered together in a bottom row, he must have freedom to say so with confidence, rather than being shaped into reciting the "right" answer. The space occupied is an important consideration in the construction of number, and any attempt to teach numbers must let the child himself work out the relationship between space and number through his own process of reasoning. If the process becomes well structured, the correct conclusion is bound to emerge. Therefore, we must work on the underlying processes, not the answer or the surface behavior.

An experiment involving children taught by S. Engelmann (Kamii and Derman, 1971) showed that in teaching language and thinking, children must be allowed to be honest with themselves. The 6-year-old children who took part in this experiment were taught to conserve weight and volume and to explain specific gravity. In the post-test, they were often found to give the correct answers, but usually in a sing-song fashion, as if they did not understand what they were saying. Teaching methods that are rooted in the child's entire cognitive framework, which is in turn rooted all the way back in sensori-motor intelligence, must encourage him to express *exactly* what he experiences and *exactly* what he believes.

In the teaching of inner-city children, much emphasis is placed on a positive self-concept and the teaching of language. But methods based on a mechanistic psychology and the perspective of the adult tend to be too direct and self-defeating. For example, a group of children had just finished drinking their juice. The teacher placed piles of cut-out circles, squares, and triangles on the table. She then gave to one child a sheet of paper on which a circle was drawn with a crayon. The child correctly responded to the teacher's request to "find a shape that's just like your circle and put it right on top of your circle." The next question the teacher asked was "Is the circle you took just as big as the one on your paper?" The child was visibly bothered because she found the teacher's circle to be about 5/16 of an inch smaller than hers all around. When the teacher repeated the question, the child said nothing and pretended not to know. The teacher then told the child that the two were the same size and went on to say, "Now, let's say it together. This (pointing) circle is just as big as this one (pointing) . . . Very good . . . Let's say it again. This circle is just as big as this one."

This is a trivial example, but it serves to illustrate how much children are taught from the adult's perspective. In attempting to teach language and size discrimination, the teacher was also unwittingly teaching other things, such as the message that the "correct" answer always comes from the teacher's head. Without theoretical clarity, good intentions can thus result in experiences that force the adult's perspective. We cannot foster a positive self-image if we do not stop to ask the child what he honestly thinks.

From the preceding comments on concrete experiences we shall turn to what is commonly called "abstraction." Piaget would agree that representation is a kind of abstraction because the child deals with objects and events that are not concrete and present. However, in his terminology and conceptualization, "abstraction" means something quite different and there are two kinds of abstraction: abstraction from the object and abstraction from coordinated actions.

Abstraction from the object takes place in physical knowledge, and abstraction from the child's coordinated actions takes place in logico-mathematical knowledge. Whether or not a needle sinks in water is physical knowledge that can be found by discovery and generalized by empirical generalization. But the notion of specific gravity cannot be discovered empirically or built by empirical generalization. Specific gravity has to be invented by abstraction from the child's own cognitive activity in combination with empirical experience. The conservation of number, too, cannot be discovered by empirical generalization.

If knowledge is simply built from the concrete to the abstract, a 4 year old who has walked all over a particular town such as Ypsilanti would know that he has been in Ypsilanti. But knowledge is not simply built from the concrete to the abstract in the sense of representation, and each child has to construct an entire cognitive structure by abstraction from objects and his own coordinated cognitive activity, so that he will have the framework within which he will be able simultaneously to understand the meaning of terms such as "New Brunswick" and "university" in a spatial, temporal, social, classificatory, seriational, and numerical sense.

From the standpoint of the child's cognitive development, there are thus many kinds of "concrete" experiences and many kinds of "abstraction." As stated in an earlier paper (Kamii, 1970), the teacher who understands these differences will have the guidelines she needs to decide what to teach and how, and what not to teach and why. We shall now elaborate on this point by discussing the role of the teacher in a Piagetian school.

THE ROLE OF THE TEACHER IN A PIAGETIAN SCHOOL

The role of the teacher in a Piagetian school is not one of transmitting ready-made knowledge to children. Her function is to help the child construct his own knowledge by guiding his experiences. In physical knowledge, for example, if the child believes that a block will sink in water, she can encourage him to prove the correctness of his statement. If he predicts that a marble placed on one side of a balance will make that side go down and the other side go up, she does not say,

"You are right," but instead says, "Let's find out." She lets the child discover the truth by letting the object give the answer.

In the logico-mathematical realm, the role of the teacher is not to impose and to reinforce the "correct" answer but to strengthen the child's own process of reasoning. For example, rather than trying to teach the conservation of number by empirical generalization, she tries to increase the child's mobility of thought in all realms—in classification, in paperfolding activities, in symbolization, in physical knowledge, etc. Only incidentally and as part of the general development of the entire cognitive structure does she ask questions such as, "Do you think there will still be enough cups for all of us after we wash them?"

The role of the teacher in a Piagetian school is an extremely difficult one because she has constantly to engage in diagnosing each child's emotional state, cognitive level, and interests by carrying a theoretical framework in her head. She also has to strike a delicate balance between exercising her authority and encouraging children to develop their own standards of moral behavior. She can much more easily follow a curriculum guide, put the children through prescribed activities, and use old techniques of discipline.

The teacher in a Piagetian school has to be a highly conscientious and resourceful professional who does not have to have standards that are enforced from the outside. The kind of teacher Piaget would like to have is the kind of adult that a Piagetian school aspires to produce—one who with strong personal standards continues to be a learner throughout his life.

In summary, the pedagogical implications of Piaget's theory suggest the kind of reform that makes learning truly active and encourages social interactions among pupils to cultivate a critical spirit. The teacher in a Piagetian school does not present ready-made knowledge and morality but, rather, provides opportunities for the child to construct his own knowledge and moral standards through his own reasoning. The emphasis of a Piagetian school is definitely on the child's own thinking and judgment, rather than on the use of correct language and adult logic.

Piaget's theory suggests needed reforms for education. Although the theory is radically different from the other theories on which current practices are based, it is also in line with the old clichés of schools of education, e.g., "the whole child," individual differences, the teaching of children to think, the encouraging of initiative and curiosity, and the desirability of intrinsic motivation.

Schools of education alone cannot change the current practices, but, by definition, it is they who have to assume the leadership in changing the state of affairs. Piaget's theory gives more precise guidelines than those that were previously available to translate the old ideals into actual practice. The responsibility of schools of education is particularly heavy because society can no longer afford to perpetuate a system of compulsory taxation and compulsory attendance in which an enormous number of consumers are literally forced to accept the services that unnecessarily create academic failures, discipline problems, and unemployable dropouts.

NOTES

1. The chapter is part of the Ypsilanti Early Education Program, which was funded under Title III of the Elementary and Secondary Education Act of 1965 (No. 67–042490). The opinions expressed herein, however, do not reflect the position or policy of the funding agency, and no official endorsement by the Office of Education should be inferred. I am grateful to Hermina Sinclair, Rheta DeVries, and Leah Adams for critically reading the chapter.

2. In the discussion that followed the presentation of this chapter, Dr. Sinclair added the following two points about the distinction between physical and logico-mathematical knowledge: (1) The significance of the distinction is that the mode of structuring is different, not just the source of knowledge. (2) Neither physical knowledge nor logico-mathematical knowledge can exist without the other. Pure logic almost exists, but

physical knowledge is involved even in the classical example of the child who always found 10 pebbles, whether he counted them from left to right, or from right to left. The fact that pebbles let themselves be ordered is an example of physical knowledge. Physical and logico-mathematical knowledge are thus almost indissociable and are seen more as a continuum than as a dichotomy.

Piaget often makes a theoretical distinction and turns around to say that the two are in reality indissociable. Assimilation and accommodation, and the figurative and operative aspects of knowledge, are other examples of theoretical distinctions that have to be kept clearly in mind in education, even though they are inseparable in reality. Awareness of the distinction makes a difference in the way we teach.

REFERENCES

Almy, M., Chittenden, E., and Miller, P. *Young children's thinking*. New York: Columbia Teachers College Press, 1966.

Chittenden, E. A. What is learned and what is taught. *Young Children 25* (1969), 12–19.

Duckworth, E. Piaget rediscovered. In R. E. Ripple and V. N. Rockcastle (eds.), *Piaget rediscovered* (a report of the Conference on Cognitive Studies and Curriculum Development). Ithaca, N.Y.: Cornell University School of Education, 1964.

Furth, H. G. *Piaget for teachers*. Englewood Cliffs, N.J.: Prentice-Hall, Inc., 1970.

Ginsburg, H., and Opper, S. *Piaget's theory of intellectual development: An introduction*. Englewood Cliffs, N.J.: Prentice-Hall, Inc., 1969.

Kamii, C. Piaget's theory and specific instruction: A response to Bereiter and Kohlberg. *Interchange 1* (1970), 33–39.

Kamii, C. An application of Piaget's theory to the conceptualization of a preschool curriculum. In R. K. Parker (ed.), *The preschool in action*. Boston: Allyn and Bacon, 1972.

Kamii, C., and Derman, L. The Engelmann approach to teaching logical thinking: Findings from the administration of some Piagetian tasks. In D. R. Green, M. P. Ford, and G. B. Flamer (eds.), *Measurement and Piaget*. New York: McGraw-Hill, 1971.

Piaget, J. Education et instruction. *Encyclopédie Francaise*. Paris: Librairie Larousse, 1935. Tome XV.

Piaget, J. *The mechanisms of perception*. New York: Basic Books, 1969; London: Routledge and Kegan Paul, 1969.

Piaget, J., and Inhelder, B. *The child's conception of space*. New York: Norton, 1967; London: Routledge and Kegan Paul, 1956.

Sinclair, H., and Kamii, C. Some implications of Piaget's theory for teaching young children, *School Review 78* (1970), 169–183.

CHAPTER 10

PIAGET'S INTERACTIONISM AND THE PROCESS OF TEACHING YOUNG CHILDREN

Constance Kamii

The background for this chapter is the work of the Ypsilanti Early Education Program, Ypsilanti, Michigan. One of the purposes of this project was the development of a preschool curriculum based on Piaget's theory. In this chapter we shall discuss the theoretical bases for the curriculum from the point of view of how Piaget sees the relationship between the environment and the young child because our educational program must be understood in comparison with other programs not so much from the standpoint of what we taught but rather how we taught and the reasons for it. We shall first consider Piaget's interactionism and constructivism; then, three general ideas drawn from his theory of teaching young children; and finally, educational implications of the interaction thesis of Piaget.[1]

Interactionism and Constructivism

In an article that quickly became a classic in preschool education, Kohlberg (1968) reviewed such diverse theories of learning as those of Locke, Rousseau, Freud, Dewey, Lorenz, Gesell, the behaviorists, Bereiter and Engelmann, and others, and grouped their theories into three categories according to how they viewed the relative contributions of the environment and the organism to cognitive development. The three are the maturationist point of view (which emphasizes the part played by the organism), the environmentalist point of view, and the interactionist point of view (which states that the organism and the environment interact in a complex and inseparable way).

There are many interactionists, but Piaget's interactionism is unique in that central to it is the concept of *cognitive structures, which the child himself constructs* in interaction with the environment in a continuous way from birth to adolescence. This interactionism and constructivism will be discussed under the following headings: (1) what Piaget means by "intelligence," (2) the biological origin of intelligence, and (3) the development of intelligence.

WHAT PIAGET MEANS BY "INTELLIGENCE"

Piaget gave his course on intelligence for the last time in 1970–1971 before his retirement from the University of Geneva. He began his first lecture by asking what is meant by "intelligence" and answered his own question by saying that for him "acts of intelligence" consist of "adaptation to new situations." He went on to say that, although intelligence enables us to adapt to new situations, situations are usually not entirely new, and we understand new situations in terms of the knowledge that we bring to them.

There are two aspects in any act of intelligence: (1) the comprehension of the situation, and (2) the invention of a solution based on how we comprehend the situation. In other words, our comprehension of the situation is always part of our adaptation to it. How do we comprehend a situation, or in a larger sense how do we comprehend each situation in our environment? We understand it by assimilating it to the knowledge that we have already built and brought to the situation. We can never see a situation as it is "out there" in the external world. We can understand it only by assimilating it to the knowledge that we have already built.

This knowledge that we bring to each situation is what Piaget calls "structures," in the sense that knowledge is always structured, or organized, in some way at all age levels. The knowledge of a baby is organized in terms of his action patterns, because he can "know" reality only in terms of motor actions and sensations. By age 2, however, the organization of knowledge begins to exist in thought, and as the years go by the child constructs increasingly richer and more elaborate structures. Because we can understand reality only by assimilating it into the structures we have already built, the environment of the 4 year old is very different from ours even when we physically exist in the same objective environment. In other words, *the very nature of our interaction with the environment* is determined by our cognitive structures.

For example, an adult standing in front of an audience seated to hear a speaker comprehends certain factors that a 4 year old standing in the same stimulus situation will not understand. Similarly, a 2 year old or a dog receives the same retinal image as a 4 year old, but the *information* they receive will be different. In the case of a fly, there is not even a retinal image, and the audience and podium are merely things to fly around or to crawl on. In other words, the information and meaning each organism receives are determined by the structures it brings to the situation.

One of the ways to study the structures into which a stimulus is assimilated is to ask different people to say the first thing that comes to mind immediately after they hear a certain word. A word that any 4 year old understands is "mommy." A 4 year old's free association might be "She loves me," or "She cooks for me." A 10 year old might say, "She stays home and takes care of the house." A 20 year old's response might be "motherhood" or some more Freudian term. The fact that the stimulus "mommy" elicits one free association from a 4 year old and another from a 10 year old illustrates Piaget's view that, because knowledge is an organized, coherent, whole structure, no really meaningful concept can exist in isolation. Each concept is supported and colored by an entire network of concepts and feelings and the "information" that a stimulus produces is determined by the developmental level of this network.

Anyone who has tried to explain to a 3 or 4 year old that mommy is grandmother's child appreciates the child's concept of "mother." The dialogue between mother and child might go as follows:

> CHILD: Mommy, why doesn't Grandma have any children?
> MOTHER: But I *am* Grandma's child. She is my mommy.
> CHILD: No, you can't be a child because you are a mommy.

No amount of explanation or audiovisual aid like mommy's baby pictures will make the child change his mind.

Piaget is often believed to view man only as a cognitive being. Actually, his position is that in reality cognition cannot be separated from affectivity. For theoretical analysis, however, he has studied cognition alone. When a person free-associates to the word "mommy" by saying "She loves me" or "motherhood," it is difficult to separate affectivity from cognitive structures.

The above examples of what the word "mommy" means to children at different age levels illustrate another point Piaget makes: When we come in contact with reality, we always transform it according to the network of concepts that we bring to the situation. The way a 4 year old transforms reality, or the stimuli in the environment, is much more static, egocentric, and deforming than the way adults transform the same stimuli. Our way corresponds much more closely to external reality because our logico-mathematical structures are more elaborate. This view that we always transform reality, or add to it, is in sharp contrast with our empiricist upbringing that taught us that the best way to achieve objectivity is to get information through our senses, like a camera taking a picture. Piaget showed that we interact with reality not as it is "out there" but as our cognitive structures transform the stimuli coming through our senses.

To summarize what Piaget means by "intelligence," acts of intelligence consist of adaptation to new situations. Whether the new situation consists of a practical problem, a theory to understand, quantities of liquid to compare, or a mathematics problem, we adapt to each situation in terms of how we understand it. Because this understanding comes through our cognitive structures that assimilate and transform sensory information, the nature of our interaction with the environment is always determined by the level of development of our cognitive structures. The next question to consider is how structures originate and develop.

THE BIOLOGICAL ORIGIN OF INTELLIGENCE

As a biologist, Piaget looked at the child's cognitive development from a biological point of view, which is very different from the way Binet looked at children's intelligence. Piaget observed that the characteristic of all living things is their tendency to adapt to their environment. For example, trees adapt to their environment. When they stop adapting, they simply die. Worms, insects, fishes, rats, dogs, and every other animal also adapt to their environment. Otherwise, they, too, die off. Animals have an enormous advantage over plants in that their biological mechanisms include the ability to move in space.

As we go higher up in the evolutionary scale, we find not only the ability to move in space but also the ability to move voluntarily on the organism's own initiative, rather than being limited to only reacting to external stimuli. Amoebas and worms only react to external stimuli and do not move on their own initiative, but higher animals like horses, chickens, cats, and monkeys can move voluntarily. This capacity to move voluntarily has enormous implications for the organism's ability to construct knowledge (e.g., the notion of objects, object permanence, representation, and all that follows).

Higher animals have the additional biological mechanisms not only to move voluntarily but also to keep moving to continue pleasurable activities and to avoid unpleasant ones. The anticipation of pleasure and pain involves considerable intelligence because indices of pleasure and pain have to be discriminated. In this context, we can view conditioning as a particular form of adaptation. All animals adapt to reward and punishment. Piaget's theory can thus be said to be not in opposition to *S-R* theory but to be a much broader theory. By referring to biological adaptation, it explains both conditioning and the construction of cognitive structures, as we shall show below.

A corollary of the higher animals' tendency to continue pleasurable activities is their tendency to play. Amoebas and worms do not play, but higher animals such as dogs, cats, and monkeys have the capacity and need to play. Through playful repetition of activities that are interesting to them, cats, for instance, have been shown to construct object permanence (Gruber et al., 1971), build knowledge about the physical nature of objects, and organize their space.

Piaget points out that the human baby has all the above adaptive mechanisms of higher animals but has a more complex central nervous system and hence greater biological potentials for intelligence. Also, the human baby lives in a social environment that provides greater protection and greater pressure on its young. Because of these differences, human intelligence develops far beyond whatever can be learned by conditioning, object permanence, and the organization of space. In *The Construction of Reality in the Child* (Piaget, 1954) and *Play, Dreams, and Imitation in Childhood* (Piaget, 1962), we see the complete continuity between the baby's acts of biological adaptation and what we usually consider "cognitive" activities (e.g., symbol formation and notions of space, time, and causality). In numerous other volumes Piaget and Inhelder describe how human intelligence goes on to develop all the way to the adolescent's ability to engage in hypothetico-deductive reasoning.

We shall now take a look at this development.

THE DEVELOPMENT OF INTELLIGENCE

One critical point about the development of intelligence is that it is a continuous process of construction from birth to adolescence in a sequence that is the same for all children in all cultures.

The environmentalist-empiricist view of learning states that the child acquires knowledge by a process similar to that of absorption, or "input," of information. Piaget, in contrast, believes that knowledge is not absorbed from the outside, but rather is constructed from the inside by the child in continuous interaction with the environment. This process of construction, as described in the many volumes by Piaget and Inhelder, is quite different from what our adult common sense leads us to believe.

If knowledge were built by mere absorption or simple input of information, it would be possible to vary the sequence of its acquisition. However, because it is built by a continuous process of construction of structures that are rooted in biological adaptation, the sequence of development is the same for all children regardless of the culture in which they live, and we cannot change the sequence or skip a step in the long process of construction.

The only major differences that have been observed are those in rate of development. Generally speaking, children in a more developed culture develop faster than those living in a less developed culture. Within the same culture, children living in the city and more advantaged groups develop faster than those living in the country and in a less advantaged socio-economic group. Piaget says that the causes of these differences are not known in a precise manner, but four factors are necessary for development: (1) biological factors, (2) experiences with physical objects, (3) social factors of interindividual coordination and cultural and educational transmission, and (4) factors of equilibration.

To summarize the above discussion of Piaget's theory, *we interact with the environment through our cognitive structures, which transform the sensory information that we receive from our environment.* The transformation is very deforming in infancy and comes to correspond more and more closely to reality as the cognitive structures become richer and more elaborate. The child goes through many stages of constructing these structures, and the sequence of this development is the same for all children.

In the simple language of analogy, when we go to a foreign country, we see more in the new environment and adapt better to it if we take with us a general framework for thinking and some general knowledge. Our task as educators is to help children to elaborate their knowledge within a rich and coherent framework, so that they will learn all through their lives and adapt well, not only to what already exists in the environment, but also to a world that we cannot even imagine and to the ideals that they will construct with their cognitive and affective structures.

The Educative Process Implied by Piaget's Interactionism

According to Piaget, four factors are necessary for cognitive development. This position suggests, first of all, the necessity of all these factors to be operative in a classroom. How to enable maturation, experiences with physical objects, social interactions, and the process of equilibration all to be present at the same time in a classroom that provides spontaneity is a large order. No cookbook curriculum can possibly give the teacher the answers in a ready-made form. The teacher must teach in such a way that all the factors are at work as she comes to know the level of each child's functioning.

A second characteristic of Piaget's interactionism is the view that knowledge is acquired by a process of construction, rather than by absorption and accumulation of information from the external world. The educational implication of this constructivism is that we cannot teach the child directly, especially in the logico-mathematical realm. When we offer an explanation to the child, or demonstrate to him, or program activities according to our adult common sense, what we think we are teaching and what the child actually learns may turn out to be two different things. The discrepancy may not be as obvious as in the case of the 3 year old who could not believe that mommy could be the grandmother's child. But the example does serve to emphasize the fact that what we teach is received by the child not directly but always through his cognitive structures. Therefore, we do better to put the accent on what the child learns than on what we think we are teaching.

For example, one day in our preschool, the teacher set up water play for the children to engage in to find out whether certain things would sink or float. One of the children predicted that a small block would sink. When she found out that it floated, she went to get a larger one, saying that that one was big enough to sink. Upon finding out that it also floated, she went to get still a larger one, predicting that this time the block would sink. During this investigation, the teacher responded with such comments as "Let's find out, Did it sink . . . ?" and (turning to two other children) "Do *you* think it will sink?" It would have been much easier for her to explain that things made of wood usually float. However, she had worked hard to encourage the children to raise their own questions and to try to answer them on their own initiative and resourcefulness. It is truly an art to teach without giving all the correct answers and without going to the other extreme of sitting back and watching children play. How to interact with children to enable them to test out their own ideas against objects and other people is an art that cannot be prescribed in a cookbook curriculum.

Of course, at the particular time described, our children did not learn the concept of specific gravity. However, they did learn something about specific gravity and were fascinated with their own research. They also kept thinking and wondering about their own ideas. For the construction of one's own knowledge, I feel that it is far better for children to wonder seriously and remain curious about the environment than to be told the answer and to learn incidentally that the answer always comes from the teacher's head.

A third characteristic of Piaget's interactionism and constructivism is the view that this construction takes place in a certain sequence that is the same for all children in all cultures. The educational implication of the universality of the developmental sequence is that, if we want learning to be permanent and solid enough to permit cognitive development throughout the child's life, we must (1) let the child go from one stage after another of being "wrong" rather than expect him to reason logically like an adult, and (2) allow for a certain slowness in the developmental process.

In a previous study (Kamii and Derman, 1971), we questioned some 6-year-old children at the University of Illinois who had been taught by S. Engelmann to answer questions that children cannot usually answer until they are about 11 years of age (e.g., the explanation of why certain objects sink and others float in water). Mr. Engelmann held that Piagetian stages are simply a matter of teaching and that the concept of specific gravity could be taught to 6-year-old children. He therefore taught specific gravity to some children in kindergarten and allowed us to come and give the post-test. What we saw was that underneath the overlay of correct answers the children had learned, their thinking clearly remained preoperational. For example, they predicted that a big candle would sink but that a tiny one would float, or that one cake of soap would sink and an identical cake of soap would float.

A Piagetian interpretation of these findings would suggest that no stage can be skipped and that development cannot be speeded up in a few weeks from the 6-year-old level to the 11-year-old level. Six-year-old children have their own way of thinking and believing. We can get surface conformity to adult reasoning, but all of us know that what we learn in this manner is forgotten as soon as the final exam is over.

The reason that it is important for us to let the child go from one stage after another of being *wrong* is that *wrong* notions usually contain a certain amount of correctness. For example, to predict whether something will sink or float, it is not entirely wrong to consider heaviness as the determining factor. This reasoning is not entirely wrong; it is only incomplete. The preoperational child needs time to construct gross structures that will later be differentiated and coordinated into more adequate structures.

We have a habit of thinking in terms of right and wrong answers and equating intelligence with the ability to pass or to fail specific test items. As Piaget points out, however, knowledge does not develop from "all wrong" to "perfectly correct." All children have *some* knowledge about whatever we try to teach them, and their knowledge always contains some elements of truth. Therefore, if we honestly want to meet preoperational children where they are, we have to figure out how *they* think and to interact with them in terms of how *they* reason.

The Piagetian philosophy of letting preoperational children

be preoperational is sometimes interpreted as a maturationist philosophy of education in which all the teacher can do is wait for "readiness" to develop. I do not draw this pedagogical principle from the theory, as the cross-cultural research indicates the considerable influence of the environment. The way in which we arrange the classroom situation for water play is an example of how within the preoperational framework we try to build "readiness" for classification around age 7–8, the concept of specific gravity around age 11–12, and a life-long attitude of research based on personal curiosity and convictions.

The fourth characteristic of Piaget's interactionism is the view that intelligence is an organized, coherent, whole structure, and not a collection of skills. The literature on the objectives of early childhood education discusses perceptual, cognitive, language, thinking, conservation, classification, seriation, addition, and even comprehension skills in reading. The image one gets from this approach is that of a machine being built by an engineer who puts many wheels together so that, when the subsystems are assembled, the whole system will function. This mechanistic model is quite different from Piaget's biological model, which views the development of intelligence as being similar to that of an embryo.

The arms, fingers, lungs, head, and eyes of am embryo develop out of a structured whole from the very beginning. If we want well-structured hands, feet, lungs, and eyes, we cannot build them separately and then put them together. The individual parts develop through a process of differentiation, coordination, and construction. This development in a biological sense is an irreversible process that takes place only in one direction. In contrast, a mechanistic process can take place in two directions (i.e., assembling, disassembling, and reassembling). In other words, a characteristic of the biological constructivist view of learning is that what has been learned once is never forgotten. For example, once the child has built the cognitive structure of number, that structure is not forgotten and remains an integral part of later structures throughout his life.

Of course, there are skills to be learned, such as reading, writing, counting, tying shoe laces, skipping, and pasting. But the important thing is that these skills be treated as tools in the service of intelligent living, not as the cause of intelligence or goals in themselves. Also, there are skills to be learned in reading, but these skills must be clearly distinguished from comprehension, which comes from the child's cognitive structures, as we saw in the different meanings of the word "mommy" at different age levels. In writing, too, there are skills to be learned, but motor skills must be distinguished from the structuring of representational space (Piaget and Inhelder, 1967). Counting, too, is a skill, but this skill must not be confused with the cognitive structure of number (Piaget, 1952).

That it is possible for us to help develop intelligence as an organized whole structure is not known for certain. One thing is certain, however: the fact that a cognitive framework cannot be built in 1 or 2 years of preschool. A longitudinal experiment is necessary to find the answer. Because no really meaningful concept can exist in isolation, it seems worth trying to find ways to develop the framework so that specific skills and information can be anchored in the total structure. If the child has a more elaborate cognitive network, he can apply it to almost every conceivable problem in such diverse areas as physics, chemistry, history, and geometry. When he has well-elaborated cognitive structures, the child can arrive at the correct answer to a variety of questions as an obvious logical necessity. This approach to preschool education seems more defensible than the alternative of trying to teach every specific skill, rule, and information in the hope that some will be remembered and transfer to other situations.

Although intelligence cannot be divided into parts, Piaget

delineates two interlocking areas of knowledge that have different modes of structuring: (1) physical knowledge, and (2) logico-mathematical knowledge. Sinclair[2] adds a third area, which she calls "social knowledge." Physical knowledge is structured from the object's reaction to the child's action on objects. Whether or not a block sinks in water is an example of physical knowledge. Logico-mathematical knowledge is structured from the child's actions themselves. Whether or not there are more blocks in one box than in another is an example of logico-mathematical knowledge. The logico-mathematical characteristic of "more" is not *in* the objects themselves but, rather, is introduced by the child through his own action. The importance of this action, of course, is not the motor act of moving things but the mental action of introducing a relationship among objects. Social knowledge, in contrast, is structured from people's reactions. The fact that blocks are for building and not for throwing is an example of social knowledge. Another example is the fact that blocks are called "blocks" and not "blobs" (Kamii, 1970).

The little girl who went from one block to another to find one that would be big enough to sink was engaged in the structuring of all three types of knowledge. The activity obviously involved physical knowledge. But, in addition, she also practiced seriation and classification (i.e., looking for an increasingly larger block and grouping things into "things that sink" and "things that float"). She also used a lot of language and understood that under the circumstances blocks could be put in water. In our work in curriculum development, we try to use this kind of play to develop children's intelligence as a structured whole.

In conclusion, the important thing about learning and teaching in a Piagetian sense is that, if the entire network of thought processes is well developed, the child will be able to comprehend all kinds of problems at a high level of intelligence. If a problem is assimilated into well-elaborated structures, the solution can be deduced as an obvious logical necessity. This point becomes increasingly clear as we see more children and study the recent research in Geneva on causality, learning, language development, and contradictions.

NOTE

1. This is a revision of a paper read at the annual conference of the National Association for the Education of Young Children in Boston, November, 1970, under the theme "Ecology: The Interrelationship between Environment and Young Children." The work described was funded under Title III, ESEA, No. 67–042490, which supported the Ypsilanti Early Education Program. H. Sinclair of the University of Geneva assisted in the development of the curriculum, and M. Denis-Prinzhorn of the University of Geneva and R. Peper of the Ypsilanti Public Schools critically read the paper and contributed many ideas.

2. Ibid.

REFERENCES

Gruber, H. E., Girgus, J. S., and Banuazizi, A. The development of object permanence in the cat. *Developmental Psychology 4* (1971), 9–15.

Kamii, C. Piaget's theory and specific instruction: A response to Bereiter and Kohlberg. *Interchange 1* (1970), 33–39.

Kamii, C., and Derman, L. The Engelmann approach to teaching logical thinking: Findings from the administration of some Piagetian tasks. In D. R. Green, M. P. Ford, and G. B. Flamer (eds.), *Measurement and Piaget*. New York: McGraw-Hill, 1971.

Kohlberg, L. Early education: A cognitive-developmental view. *Child Development, 39* (1968), 1013–1062.

Piaget, J. *The child's conception of number*. New York: Humanities Press, 1952; London: Routledge and Kegan Paul, 1952.

Piaget, J. *The construction of reality in the child*. New York: Basic Books, 1954; London: Routledge and Kegan Paul, 1955.

Piaget, J. *Play, dreams, and imitation in childhood*. New York: Norton, 1962; London: Routledge and Kegan Paul, 1951.

Piaget, J., and Inhelder, B. *The child's conception of space*. New York: Norton, 1967; London: Routledge and Kegan Paul, 1956.

CHAPTER 11

THE USE OF CLINICAL AND COGNITIVE INFORMATION IN THE CLASSROOM

Mireille de Meuron

The work of Piaget (1959, 1966b), Inhelder (1963), and their collaborators, as well as related literature in England, the United States, and Canada has supplied us with a wealth of facts and observations on children's intellectual development. In the light of such knowledge, certain traditional aspects of our education have been called into question. Indeed, information on child development has not been and is not now circulating as much as would be desirable or even necessary among professions dealing with children. This fact is relevant not only to teachers, pediatricians, clinical psychologists, and social workers, but to the writers of curricula and textbooks, as well. The complexity and quantity of this information regarding the child's growth of thinking together with the language in which the information is delivered tend to prevent new knowledge from becoming readily accessible to the various helping professions.

In education, this lack of communication has become particularly dramatic. On the one hand, change in schools has become a political issue and, on the other, the extent of school failure by many children for whom it was the only way of escape has alarmed teachers, school officials, and resource services. Consequently, there has occurred a large consensus on the need for change, if not for a specific type of change.

A number of educational theorists have interpreted the problem within their respective frame of reference and have developed such approaches as computerized teaching equipment and programs, elaborate audio-visual aids, reinforcement techniques, perceptual exercises, and the like. Notwithstanding the justified desire for rapid solutions to this uncomfortable problem of school failure, no panacea exists. No series of techniques, kits, or gadgets will bring about satisfactory results even if these aids are very good. Dramatic difficulties impede the school life of a great number of children because the schooling and teaching they receive are not compatible with their needs, aptitudes, and idiosyncrasies. We now widely recognize that actual change of teacher effectiveness in the classroom will not be brought about simply by the promulgation of laws or the spending of large sums of money. Although some of these may be useful in setting up the conditions allowing for change, they will not touch the process of that change and particularly the transactions at the reality level in the public school classroom. What is required is an adaptation of teaching to the children who are taught and this process is always a complicated one. The teacher must be able to offer learning experiences that permit the child to utilize and to exercise his particular level of thinking in interacting with certain features of the task or activity in which he is engaged. This process will be the focus of the following section.

Information on child development constitutes both a sound longitudinal rationale for the creation of new methods of teaching and a powerful impetus for moving the change process ahead. In fact, such information implies not only a change of content but also a modification of the whole conceptual canvas of interpretation and, therefore, of decision-making. A teacher who knows how children think will modify her perception and understanding of what is taking place in her classroom and, consequently, the quality and goals of her interventions. In the long run, the teacher can learn to respect and to use the idiosyncrasies of the level of thought considered. Of course, this achievement requires that such information, accepted and understood, is catalyzed within the school for the benefit of change. In a context of the school difficulties of a child, we have noted, for example, that child behaviors are frequently interpreted inaccurately as bad will (lack of attention or persistence, poor memory, perceptual defects, etc.), or more perniciously attributed to an organic cause such as retardation or brain damage. If these labels are removed and behaviors are viewed in their developmental perspective, they may be dealt with in a more positive fashion. This approach in itself will not solve the problem, but it may bring about a reconsideration of whether it is essential or economical to face certain children with learning tasks that may be inappropriate or too difficult for them. Knowledge of the child can also raise a question about the method and modality used to present the task and whether these are best suited to the logical tools that the child is known to use in order to understand it. This knowledge also makes possible the formation of helping measures that can be appropriately selected and combined and the invention of new measures, as well.

Knowledge of the direction toward which the child's thought is evolving—knowledge of the dialectic of this evolution or part of it—is in itself a force for change. It provides the framework for eliminating teaching approaches that constitute obvious, useless obstacles that go against the stream of development. It makes possible creation of a new methodology that would use the children's mentality and adapt classroom composition and management to it. It fosters the development and introduction in the curriculum of a number of activities directed toward the development of thought. Although research has not been conclusive in regard to using a controlled way to teach children to think, some factors contributing to intellectual development have been singled out. Their possibilities should be explored and used. Much basic research still has to be carried on not only to create curricula but to set up a methodology of evaluation of the results obtained.

Initiation of Classroom Changes

To accomplish changes in a classroom setting requires a constant translation of information from basic research into the educator's language and careful experimentation with its implications in a practical vein. Unfortunately, theory sometimes suffers in the process. The examples given in the remaining sections of this chapter will be drawn from an attempt to introduce such changes in a New York City public school. The project came about as a cooperative endeavor of a school district administration and a nearby community clinic. The Gouverneur Health Services Programs provided biological, psychological, and social health care to a district housing 144,000 persons (100,000 of whom were believed to be medically indigent). It welcomed participation of systems important in the life of the families it served. The schools referred children for help with their school problems and were generally expected to participate actively in the planning and carrying out of treatment.

An exploratory study of children referred to a clinic in the Behavioral Sciences Department at Gouverneur preceded a pilot classroom project. The favorable response from the school district administration toward the classroom project resulted in an extended project that added a number of classrooms. In the ensuing discussion references will be made to three groups: (1) the Referred Group seen at the clinic consisting of 57 children ranging in age from 6 to 14, and enrolled in kindergarten through fifth grade; (2) the Pilot Classroom

Project consisting of children and their teachers in 4 prekindergarten classes and 1 kindergarten class in 3 different schools; and (3) the Extended Classroom Project consisting of children and their teachers in 10 classes including prekindergarten, kindergarten, and first and second grade. An average of 20 children were in each classroom. Three teachers were involved in the Pilot Classroom Project, and 11 in the Extended Classroom Project.

In this impoverished section of New York City, just as in many large urban areas throughout the country, the situation described by some as the "drop-out conveyor belt" has reached proportions where it is punishing all concerned. For the teacher it is often an harassing, disappointing experience. For the children it is particularly tragic because schooling has the potential for escape from poverty and the life of the slums. The cycle starts with unadaptive behavior in the child at entrance to school, followed by nonsatisfying achievement, failure, temporary exclusion, and finally referral to agencies outside the school for solutions. If one studies these failures, one is struck by the repetition of a similar constellation of factors and characteristics, even if the label arrived at differs depending on the diagnostic perspective of the professional consulted. When one observes children of this area in the regular school classrooms, one often notices the same array of problems as the children who are referred, only in a milder form. Often the teacher takes advantage of a school visit concerning a referred child to seek advice about several other children who demand and receive less attention to their needs but experience the same kinds of difficulty. The degree of the acting-out behavior in the children and the individual tolerance of the teacher determine to a large extent which children will be referred out of the system for professional help. This fact has caused many to doubt the value of a diagnosis developed out of a context of illness and pathology. De Meuron and Auerswald (1969) point out that it is obvious that the children who habitually disrupt their classrooms are not all psychopaths, psychotics, retardates, or brain damaged. The escalation in label drasticity contributes principally to worsening the response of the environment toward these children.

A view considering the ecology and the sociology of this population has been a recent development in the work of such investigators as Minuchin (1964), Auerswald (1965), and others. In the light of their findings we began study of the cognitive profiles of the children. We observed and tested routinely with Piagetian types of tasks all cases referred for socially dissonant behavior. Our intention was to find out the extent to which the children had developed the cognitive tools needed to identify, classify, and integrate the messages they received from the various complex systems in which they were expected to function.

We shall briefly describe the group studied so as to make clear the directions of the change we subsequently sought to bring about in our classroom project.

Affective and Cognitive Characteristics of the Children in the Referred Group

The majority of the 57 children selected for this study were between 7 and 11 years of age. We eliminated children obviously deeply retarded, those with severe speech problems, and those with long histories of mental illness. Our data were not collected within the framework of formal research and therefore reflect the numerous encroachments due to service imperatives. Most of the standard psychiatric categories were present: borderline retardation, hyperactivity, brain damage, minimal brain dysfunction, behavior problems, emotionally disturbed, and schizoid. All children had serious problems in school both in achievement and behavior. Most came to our attention as a result of a referral initiated by the school.

This group was remarkably diverse ethnically as is the population from which it comes: The largest percentage was Puerto Rican and black American. Other children came from Chinese as well as white American backgrounds, namely Jewish, Polish, and Italian.

Behavior (usually described as "bad") headed the teacher descriptions of children referred. The child "cannot sit still, disturbs the classroom, acts out." His learning seemed to be impaired by the lack of focused attention and by a constant search for contact with the teacher. His vocabulary was limited and many concepts such as time and space were totally inadequate to learning in general and reading in particular. Mathematical skills usually were less impaired until the higher grades where reading problems begin to interfere.

Of course, many variations existed. The social worker nearly always reported substandard living conditions. The family was preoccupied primarily with survival and was often disorganized. The present members of the family varied (father, mother, grandmother present or not), but its socialization processes remained similar. Family patterns differed markedly from those of organized families. Minuchin et al. (1967) have noted that the children of such families were not being usefully equipped to face the school experience. Child behavior in the family was controlled by global directions such as "Quit it," "Get out," or "Stop," directions so undifferentiated that the child could not learn from the adults the rules by which to regulate his behavior or even which part of his behavior was not acceptable. He did learn that behavior was related to the person who exercised power and thus would tend to organize his environment in seeking controlling contact from the mother. Power transactions represented a large part of the family interactions. The children did not learn to elaborate a discussion or to prove their point by the choice of logical argumentation because they knew they would not be heard and that the content of the topic would be switched very rapidly. Skills for apprehending, ordering, and communicating information were not developed. Observers mentioned the important amount of communication that was carried by body movements, gestures, and facial and vocal expressions.

The child from such a background enters school with a crucial handicap. Nearly all information in the classroom is communicated verbally. Behavior at school is supposed to be controlled through rules. Some of these rules are explicit; many are implicit and constitute expectations rather than rules. For example, conflicts between children are required by school (and middle-class practices) to be resolved by discussion or teacher intervention, not by fists. This environment is extremely disconcerting for the child who cannot conceptualize the components of school rules and is not prepared to deal with them. He is faced with increasing internal disorganization and resulting anxiety reaches a very high level. He acts out his distress, thus provoking the only type of organization known to him, that is, the inhibiting control of the adult. Auerswald (1968) describes the difficulty these children have in identifying affective messages and nuancing their responses to such messages. Very few different categories are available, whether "hate messages" or "love messages" are concerned. A whole range of graduated responses may not be mobilized appropriately and, of course, the child is not in possession of the language designating such messages. This fact is very often visible in the resolution of conflicts in school where direct, intense, even violent aggression is triggered by what are often very minor causes.

The lack of differentiation of children's understanding, visible in many classroom episodes, instigated our in-detail study of the cognition in such children, not only its operational capacities but also its classification schemas. Regardless of earlier diagnostic labels, all the children in the Referred Group who came to us were studied with a focus on the ability to classify and to conserve in order for us to gather some clues regarding their possible idiosyncrasies. The interviews systematically included the following tasks: conservation of number, substances, and liquids in an adaptation as close as possible to Piaget's original techniques; seriations and classifications; and inclusion tasks and quantification of inclusion. In all cases, the interviewer made all possible effort to ensure that children understood what was expected of them. Almy et al. (1964) had established a 1-year delay in the acquisition of the conservation of number (low in the hierarchy) for children from the same area. This delay lengthened to approximately 3 years with our Referred Group. This late stabilization of operational equilibrium means that many children attending second and third grades and some in even higher grades will attempt to grasp the world and particularly school information with a thought organization still intuitive and dominated by perception, not reason—a thought organization resembling that of younger children.

To be more specific, the children's reasoning proceeded from one particular instance to another without relating it to the whole. It was transductive, not yet deductive. The child would jump to conclusions without going through all the necessary steps of the deduction. We had observed this to be a very serious handicap in attempting even very simple problems of arithmetic. Teachers complained that the children "write whatever comes through their mind" in answering the questions. This "whatever" came from the child's internal logic, a logic that the teacher needed to learn to understand. Piaget (1966a) described intuitive thought as being syncretic. Elements are linked together by the child in an immediate global fashion instead of in the logical structures of the elements because the child is unaware such elements may fall in the same category. This fusion between elements was observable in many instances in this group of children in their way of handling affective messages. They employed egocentric thought that was juxtapositional, that is to say, regardless of causal relationships one element would be placed side by side with another. Verbalizations were joined by *and*, *and*, *and*; very seldom did we hear *since* or *because*.

The reader of Piaget's descriptions of intuitive thought, egocentric mentality, and language may easily see a confirmation of these characteristics when observing children from backgrounds similar to the ones described here, children in school at an age when ordinarily this stage should be replaced by a more concrete type of organization of thought. The fact that thought remains so close to a "mental experiment" for a longer period of time perhaps explains the poor success of so many classroom lessons. Even if actions are not actually performed, representations in thought are very close to action. Mental processes proceed from configuration to configuration in a very static, concrete, irreversible manner.

Our present way of teaching, largely verbally from first grade upward, does not allow a child who is still preoperational to learn through action on objects. In kindergarten, stress has already been put on words and perception. In first grade, our child cannot rely even on the memory of the configurations now referred to in words. The overwhelming use of symbols and deductive processes in teaching does not allow the young child, still thinking prelogically, to explore mentally the situations taught. According to his abilities and motivation he will therefore either learn by rote or lose interest and misbehave. Using objects and materials in a demonstration by the teacher does not replace providing the child with opportunity for effective action and discovery of the effect of this action by the child himself. Preoperational thought, characteristic of a normally developing child in the years 2 to 7, is egocentric in the sense that it centers on one single feature of the object at a time, thus neglecting other important compensating aspects. Thus, multirelational materials must be available for action for quite a length of time so that several features can be organized simultaneously by the child and a variety of schemas developed.

Studying classification processes in our Referred Group, we did indeed confirm and underline this picture of delay in thought development. Only one-third of our 11-year-old children (about 15 children), when given an array and asked to "put together what goes together"—a task that requires them to identify and use one criterion (as shape, color, size, or function)—could shift from the original criterion they used to a different criterion, a feat we would normally expect 6 or 7 year olds to be able to perform with ease. Even when the classification was started for them, the children continued to use that new criterion for one or two pieces, and then returned to their initial criterion. Understanding correct relations of quantity between classes and the whole was extremely rare. Most children attempted classifications by constituting figural collections of the items, that is, picture-like arrangements, rather than being able to group. Some children who in the end succeeded in shifting from one criterion to a second, first went through the entire series of classification schemes described by Piaget. The inability of the children throughout these tasks to discriminate the similarities and differences between objects and situations was striking.

The attitudes of children toward even the most simple problem-solving also showed a lack of confidence in themselves and in their ability to reason. The children gave few spontaneous verbalizations. Judgments from them had to be elicited by questions formulated in such a way that a "yes" or "no" answer was possible. Almost never did a child offer a spontaneous justification of the reason for his performing a task a particular way. He rarely used the word "because" by itself when questioned; even more rarely did he use it followed by an actual reason. Even when a good contrast was established and after unlimited manipulation of the material, children would try to rely on their memory rather than to try to figure out the task. Answers such as "I don't know," or "In my school we haven't learned about juice yet," or gestures indicating a lack of understanding were the most frequent reactions.

We found that these behaviors were extremely widespread regardless of how easy the question. Children behaved as if all the situations had to be mechanically learned. In fact, one of their most frequent strategies was guessing or avoiding a choice. Manipulating data, reasoning, and figuring out the answers were never a spontaneous attitude. It seemed to us that several factors contributed to the presence of this attitude. The children were faced with information given in a form that their schemas and preoperations could not integrate. In addition, its content was often foreign to the culture and environment of the children, and nearly always to its language. But the problem was compounded by the emphasis put on memorization and the importance attached to accumulation of information in the school. This seems particularly true in schools for lower-class children. They are rewarded when they give the right word or the correct answer. Reasoning, per se, is not rewarded. As a result the answer becomes all important. Such an expectation can only increase the tendency to global responses and the automatized reaction of the child, who begins to expect automatically that he does not know. We must mention in this context the extraordinary extent of magical thinking in the reasoning of these children and in the adults in their families as well. Cause-to-effect relationships remain very obscure and often distorted until quite an advanced age. This lack of awareness of contradiction is particularly misunderstood and misinterpreted by teachers, often being attributed to stupidity in the family rather than to ignorance as a function of a lack of opportunity to learn.

In brief, we can say that all the characteristics described by Piaget as prelogical thought applied to our Referred Group sample, but they lasted for a much longer period of time than they did for the majority of their schoolmates. Tests of conservation as well as classification, together with numerous observations of children in both work and play activities, showed that the children in the regular classes, not referred, shared many of the features described above, but to a lesser degree. Seminars with teachers also confirmed the difficulties of children from disorganized, poverty backgrounds to be similar to those of children actually referred.

Working with Teachers in the Pilot Classroom and Extended Classroom Project

We proceeded to work with a group of teachers using Piaget's theories and observations cited previously as a theoretical frame of reference. The work of American authors such as Elkind (1961), Almy et al. (1964), and Sigel (1964) was used extensively. These experimental studies were referred to when relevant, and some of the tasks, wording, and ratings were adapted. Indeed, these papers among others constituted much of the basis of the original literature distributed to the teachers.

Our project attempted primarily to bring directly into the classroom information coming from all the ecological systems studied by the diagnosis and treatment team. Information on cognition was considered essential because it affected the children's capacity to understand and to manage their environment. Information from studies of family styles of socialization and from more psychodynamically oriented research was also brought in. Some children's test results were synthesized and interpreted with the teacher in an attempt to allow teachers to use all the facts and clues, affective and cognitive, gathered by the psychologist and his team. This information conveyed to the teacher did allow some of the referred children to remain in school. More importantly, if the teacher understood these children, she might be able to create a milieu more fertile for the less stigmatized peers of the referred children. To bring this information to discussion and especially decision-making about changes and conclusions by teachers raises many conflicts even in the most open circumstances. Some difficulties stem from tradition, some from insecurity, most from the fact that many people in the school system have become very defensive under a variety of pressures. Many ideas could not be undertaken in our project because of resistances, but we describe here changes that we were able to bring into focus in 3 kindergartens and, to a lesser degree, in some first grades and 1 second grade. All changes across all 3 grades had as a general goal the fostering, encouraging, and development of children's thinking. Teaching was adapted to the actual level of children's thinking regardless of their age.

We first observed all the classroom groups in order to get a general idea of symbolic and operational levels. Children appearing to function above or under class average were administered a battery of tasks drawn from Piaget's work. These test sessions were used to begin to acquaint teachers with Piaget's theory. Some children were chosen for follow-up with the intention of being studied in more detail by teacher and psychologist together. This was a particularly effective way of shifting teacher's thought from results and scores toward a more process-focused orientation.

Teachers thus had the chance to discover logical reasoning across several stages and start formulating for themselves the questions of the implications of cognitive development for teaching, for learning, for curriculum planning, and for classroom management. We aimed to institute some of the changes because of their specific relevance to our population.

Changing Classroom Procedures

The changes we considered together with the teachers involved three features that are interrelated, but that we shall differentiate here for clarity's sake.

The first aspect refers to changes intended to adjust teaching to the characteristics of the stages or substages represented in the classroom. This means, firstly, removing the obstacles that the present way of teaching presents for the thought organization or level of a child, and, secondly, developing and enriching the curriculum of activities selected to develop judgment. In other words, this change required shifting the emphasis from the teacher to the learner. It permitted the child to learn from doing and actions rather than from teacher explanation and verbal imitation. It allowed and fostered the multiplication in the child of schemas and their integration (i.e., ordering, classifying, identifying, etc.) rather than the accumulation of multisensory stimuli and labels.

The second aspect refers to changes involving the very delicate classroom transactions, their quality, and orientation. Modification in teacher's intervention depends upon what she is trying to bring about in the classroom; therefore, understanding the theory of development is necessary. Ideas such as those of equilibration and regulations are essential but their implications are difficult to grasp at first. In each activity and throughout the day, many opportunities arose to challenge the equilibrium reached previously by the child and to bring about new regulations. By pointing out conflict (or allowing it to emerge from the group) and by encouraging decentrations of all kind, the teacher helped to find different strategies. These were attempts to bring into the child's awareness emerging coordinations of his actions, connections, and processes. At first this was very difficult for the teacher because she had to refrain from teaching—from reinforcing and rewarding the correct answer or from correcting an answer that may have been incorrect to an adult but appropriate to the level of thinking of the child.

The third feature points to changes concerning the materials and their techniques of use. Allowing children in the same classroom to function at different levels, and the same child at different levels, successively supposes an environment with rich opportunities and with a low competitive level. Many group projects were developed in which children of different stages and with different interests could find a basis for constructive play and work (Nuffield Mathematics Project, 1968).

When teachers became familiar with Piaget's theory, the questions of the place of verbal ability in teaching arose very rapidly and constituted at first the cornerstone of discussions. The puzzling similitude of the responses from children regarded as dull with those of some of the bright ones launched the first questions by the teacher. It seemed that verbal expression was often confused with thought itself. The illusion that concept formation developed through the teaching of words was extremely deep rooted. It was, of course, much easier to see the relationship between reading and language than between reading and classification. Other exploratory and manipulative activities tended to be regarded as "play" in the condescending attitude of some teachers, or more positively, as "reading readiness." Reading was the school's greatest concern and consequently diverting time for activities seen as play was largely perceived as a waste and was strongly opposed.

Piaget (1955) has studied the relationship between language and thought. More recently, Sinclair-de-Zwart's (1968) study clarified the interdependence existing between language and operations. It is indeed crucial for children to learn words and sentences not only to label classes but to designate their actions upon reality such as comparing, compensating, etc. Teaching such words in the absence of the child's understanding and logical organization of the idea will in no way cause these understandings to emerge from the repetition.

Piaget's work on the construction of reality (1954) and on play and symbolization (1962) shows very clearly the process from action to operations, a step-by-step construction of action schemes, their assimilation of and accommodation to the object, their gradual internalization, and the onset of symbolization. In *Traité de Psychologie Expérimentale* (Fraisse and Piaget, 1953), Piaget reminds us that, if some types of notions are drawn from experience, and it is undoubtedly the case at the preoperational level, it is not from mere physical form of experience but from a logico-mathematical experience specifically. Notions also come from acting on objects. However, the new knowledge results not from an abstraction from the object as such but rather from an abstraction from the subject's action on the object. If this is the way thought progresses, then we can rearrange classroom materials, groups, and time so as to provide opportunties for such a process to take place and develop, and can intervene so as to bring into focus the cognitive processes pertaining to any activity in which the child is engaged.

The Classroom Store as a Basis for Developing Reasoning

Many kindergarten teachers have the children build up a store in a corner of the classroom. Concepts of size, shapes, and the notion of exchange constitute the present purpose of this activity. Teachers emphasize verbalization and concept formation by comments and questions such as, "Give me the *large* jar of mayonnaise, the *small* can of coffee," etc. In contrast, commenting, "You have *rearranged* your store today?" or "How come the coffee can is with the milk?" and engaging children in a discussion of the basis for their organization, their reasons, and the process they used help to bring classification schemes to the center of the whole group of transactions. In such instances, the teacher must absolutely avoid teaching in the traditional sense of the term. She must listen and make the effort to understand. Child thought is extremely varied, and it is essential for the teacher to "waste the time" to find out which path a child's thought is following. Otherwise, there are many chances that the teacher's intervention will be useless or will even stop the process trying to emerge. To the question above, "Why is coffee stored next to the milk?" the answer obtained was, "Because they are both used for breakfast!" To find this out implies that the teacher did not teach "coffee belongs with the cans because it comes in a can," or "milk should go in the refrigerator department." Such discoveries the child will make later as he reorganizes, observes, and discovers for himself. Reciprocal exchange is essential to the development of socialization and to the education of focused attention. For a child to learn what it means to pay attention, he must have experience being paid attention to.

Quite some time evolved before the child could express verbally the reasoning underlying his action, especially if the group involved were homogeneous cognitively. We found that some heterogeneous or multiage grouping allowed a much richer experience, although it generally required more adult attention. Often altercations resulted from the fact that certain types of strategies (basis for classification in the store) seemed more "right" at one level of thinking than another. Some children saw a necessity to fill up the space on one shelf before starting another collection on the next shelf. Another group might throw all the items on the floor and classify according to color, size, or type of container. Most first graders tried to replicate the usual department store organization. If the cognitive levels in the group were not too far apart, the discussion could be extremely fertile, not only because each child had to become aware of his reasoning and to stand for his point of view but also because, since his thought is egocentric, the aspect he had actually explored was privileged in his view and he had the tendency to persist in it in a very static fashion. This equilibrium (satisfaction with his own reason) might be challenged by that of his companion (who has a different reason) and reciprocal interactions occurred.

The teacher's intervention here was delicate but extremely interesting. Many different possibilities were open according to the time of the year and the evolution of the group. She might have helped children to discover and to apply new classification schemas. She might have oriented the group toward seriation and ordering of the items in the variety of ways permitted by the material. A first-grade teacher chose to bring up the idea of exchange and introduced the notion of quantities in the operations involved in all the exchange transactions. Thus, mathematics was attempted first in action and then in symbols. All sorts of existing logical materials were very useful. More classical but necessary information could also be fit in, such as writing (labels, shopping lists, sales tickets, etc.). In kindergarten one of our groups made money by painting symbols of what they wanted to buy. Successively different levels of symbols were explored. The children sometimes forgot the meaning they had given to their pictures and tried modifying the system so that they could all know what they meant!

Taking the clues from the children's actions and helping them develop to peak potential were the essential tasks of the teacher and also the most difficult to acquire when one had been accustomed to "teach at a class." At the beginning children were somewhat thrown out of balance by the freedom to act and by the lesser degree of teacher's direction. This the teacher compensated for by giving very flexible kinds of suggestions adapted to the group, such as "Please organize that store" or "Straighten up the store!" A little later a discussion of what the children wanted to achieve took place with the teacher so that the group was not left entirely to itself and such discussion also allowed the teacher to take good note of the evolution of the thinking of the children.

The Classroom Store as the Basis for Learning to Handle Conflicts

Teacher's interventions may also deal with the learning of socialized resolution of conflicts. When altercations arose, the opportunity was excellent to introduce notions of reciprocity, bargaining, and rules. The teacher did not step in immediately and arbitrate who was right. She had observed the group and found out the kind of give and take of which it was capable. In our example above, for instance, the solution might be situated at the action level—each child in turn executed his idea. Or, on a much more cooperative basis (in the sense of doing together), the group was given the task to find out in how many different ways it could order the store. First graders were asked to chart the ways on paper. This task gave place to a very fertile vocabulary lesson wherein children asked for the words they needed in the context of what they were building.

It was important that conflict was handled so as to keep the group manageable, but in the particular case of the population we have described the task was essential in terms of the mental health of the children. Conflicts had to be allowed to arise so that the children learned socially acceptable ways of resolving them. In our experience, the fear held by new teachers because of the amount of primary aggression in their class was very detrimental to the group members, who sensed it immediately, of course did not understand it, and were apprehensive about how to cope with it. The meaning of conflict to children, as we have seen from studies of the families, was complex. The rules and ways of resolving conflicts had to be taught through the teacher's understanding and guidance just as deliberately as any other subject, for if such rules were not learned at school they were learned not at all by the children or only in a punishment context. Although a certain amount of controlling interventions by the teacher always took place, in the younger grade the clear explicitation of all attitudes, expectations, and rules was essential. We were dealing with the type of children who had to make a great effort to understand what we explicitly said was expected. The children's confusion was complete when we worked according to a more or less "hidden agenda," even if the adult's goal was to try to make things easier. The preferable procedures were to consider which kind of activity was about to take place and which conditions in the group were required so that the activity might evolve. The rules that would create such conditions needed to be worked out, explained, and established. The teacher had to be convinced in this matter that the child, even if well behaved, was only in possession of very crude images of the milieu of the school. He usually had no idea of the specific expectations taken for granted in a different milieu. He has been repeatedly warned, "Be good in school or I'll kill you!" He therefore did not know either what behaviors were considered good in school or what the punishment would be if he was not "good."

It took some time for our teachers to find out which group combinations were most productive and which led to apathetic or explosive outcomes. Many factors interplayed here—cognitive level, temperament, ability to communicate, interests, and fatigue level of individual children, to name a few.

However, most of our small group of teachers soon developed a sense of "who can bring what" to the group through observation of the children. Of course, what we undertook here required a rearrangement of time allotments in the day. Each group chose a different activity and the teacher had to be able to go from group to group to help constructive action develop among the children. When discussions were underway, she did not want to have to interrupt because of class routines. Often school routines created short time slots. We attempted there, also, an adaptation to children's own rhythm. We had seen that our children had a very short focused attention span. If we broke the flow of attention when it was directed to an activity, we did not reward this involvement and actually taught the opposite attitude. An involved group was left to perform the routine independently later. Inversely, children who did not as yet function in a group were directed more. Routines were useful to their sense of security and economical of teacher's time. They were established in a way that permitted children not to be interrupted, while others were not kept waiting. In terms of learning, the teacher's intervention oriented activities to different areas. The examples given before were developed in that perspective.

Some Comments on Utilization of Concrete Materials

Materials can be very useful, of course. They allow children to anticipate their actions upon them, to act out the transformation, and to check the results produced by their actions against their anticipations. Materials allow simple concrete parts of the above strategy to take place. The teacher's role is then to help children make understandings explicit and to bring them into awareness. We have seen, however, that it is not from the material itself that knowledge will be abstracted. Materials allowing for a great variety of experimentations by the child are numerous. In our experience not too many but rather a few very flexible ones should be used at first. Teachers will know the range of children's behaviors and intuitions more thoroughly. A basin of Kosher salt was used in one of our classrooms. It has the advantage of dissolving with water and being less messy than sand. A variety of differently shaped containers, conventional measuring cups, and roles of paper designed for adding machines constituted valuable supplies. These offered a multitude of opportunities for succession of actions, recognition of similarities of results, transformations, etc., to be acted upon and the schemas corresponding to them to be progressively constructed and interconnected.

The multiplication of gadgets, kits, and toys in classrooms comes from a belief that one learns from the mobilization of the senses. Although some are the source of interesting information and should be valued for this, it is the relational aspects of materials that should attract our attention. The variety of action schemes applicable to the materials, the extent to which the effects of the action are visible, and the ease with which new understandings may be effectuated are important features of the materials to choose. Those that allow many different projects and combinations to evolve during repeated experiences over a period of time—such as blocks, clay, tinkertoys, sand, and water—are interesting for it happens often that a child seems to manipulate the situation in a very haphazard way for some time. Then, suddenly he begins introducing very sophisticated types of organizations and projects.

In order to clarify the difference in abstraction constructed by the child from his actions upon the object in contrast to abstraction derived from the physical qualities of the object, we shall give the example of one work unit and its transformation.

A Seasonal Learning Unit

A very experienced teacher had carefully planned a series of work units all concerning the fall season. Among the fruits of fall, apples were selected for the focus of an afternoon's work; all the sensory aspects of the object—shape, color, smell, taste—were to be discovered and verbalized by the children. Emphasis was placed on words and on the instruments people use when handling apples, peelers, and knives. Each child was to peel his apple, cut it, and eat part of it. The remainder was to be cooked and mashed to make applesauce. A whole science unit was planned for the following day: Heat, its effects on food, geography, the origin of the apples, transportation to the city, etc., were to be included.

In implementing the plan on the first day, one child had completed peeling the apple and had cut it into several bite-size pieces. She was teasing her neighbor saying, "I have a whole lot more than you." Interrupting them, the teacher said, "Stop teasing and tell me how you think the apple tastes," thus missing the chance for helping the two children compare number, amount, and size. This observation stimulated the beginning of many teacher discussions on conservation of quantity, liquids, and number.

Although many of the features of the Fall Learning Unit were retained, their presentation technique was modified. Consequently, a whole complementary work unit was organized: part-whole relationships, quantities, exploration of transformations resulting from the children's actions of cutting in a few or greater number of sections, reassembling the parts to make a whole, cooking and mashing and recognizing that the apple could not be reconstituted (transformed "back"). The teacher let the discussion among the children evolve as to whether, after one cut of the apple (in half), there was more, less, or the same amount as before. She asked the children to anticipate what might result from their pending action and then let each child act, observe, and discuss. Her explaining to them that repeated cutting produces a large number of apple parts and not more apples would have stopped the discussion. Some children would have come up with right answers, some with wrong. In no way would this have encouraged them to use their assimilative schema and let its equilibrium be challenged.

Further construction of these basic elements of understanding will not be altered by the teacher's explanation. At best only a response will be learned with the accompanying message that somehow magically some people know and others don't know. On the contrary, when each child acts out his cut, some will respond operationally that there is the same amount of apple, others will reason preoperationally that multiple pieces (section of the apple) are more than one piece (the whole apple). The discussion among the children can be stimulated by questions such as, "What makes you think that you have more?" or the children may be helped by reviewing in imagination the transforming action to recall the initial state, "Make believe that we put all the pieces together again. Then how much would we have?" Mobility of reasoning, not a learned right answer, is the goal. Part-whole relationships can be further explored through other different modalities—for example, jumping a given distance with different size jumps, painting or cutting paper, both of which focus on the effects of action.

In kindergarten, many of the subjects planned by teachers need not be changed as topics except that the purpose is different. Making Christmas trees is not taught to learn Christmas tree-making but may be used to calculate, compare, and introduce relationships. Such understandings are at the core of all the interventions, questions, and explanations from the teacher.

The Challenge of Change

In conclusion, we must point out, the enormous research and control efforts that must be accomplished if such an approach is extended to higher grades. There a more structured curriculum is necessary in order to inform without lacunas and to allow a certain mobility from school to school. Such planning will lead to failure, however, if it does not allow adaptation of curriculum to both the present cognitive levels and the potential cognitive abilities of the group.

Several problems pertaining to change in the present school system are equally crucial. These will have to be faced and resolved. Even if we do prepare students adequately in teacher's colleges, this achievement is endangered when the first-year teacher adapts to certain schools. A certain amount of change has to become real in the school as it exists now. From the experimental project carried out in sheltered situations we have to learn ideas, techniques, and systems. However, we also must learn how to introduce change in large systems. A methodology for change does not exist yet. Very little literature is helpful when one wants to learn the techniques, dynamics, and pitfalls of being a "change agent." As a result many such professionals are neutralized in their effort to change the situation and let political or community groups carry on most of the task and force of change. These groups are not usually in possession of scientific content information. Such a dichotomy becomes extremely wasteful. When microcosms are created, this vicious circle is resolved through the possible alliance and communication of individuals. Elsewhere it remains the major obstacle to change and to the development of creativity in matters of education.

REFERENCES

Almy, M., Chittenden, E., and Miller, P. *Young children's thinking*. New York: Columbia Teachers College Press, 1964.

Auerswald, E. H. The role of social isolation and disordered cognitive development in the genesis of dyssocial behavior, non-medical drug usage, and learning disorders in children and adolescents. Paper presented at the Seventh Western Divisional Meeting of American Psychiatry Association, Honolulu, 1965.

Auerswald, E. H. Cognitive development and psychopathology in the urban environment. In Graubard, (ed.), *Children against school: Education of the delinquent, disturbed, disrupted*. Chicago: Follett, 1968.

De Meuron, M., and Auerswald, E. H. Cognition and social adaptation. *American Journal of Orthopsychiatry* 39, 1 (1969), 56–67.

Elkind, D. Children's discovery of the conservation of mass, weight, and volume: Piaget replication study. *Journal of Genetic Psychology* 98 (1961), 219–227.

Fraisse, P., and Piaget, J. *Traité de psychologie expérimentale*. Paris: Presses Universitaires de France, 1953. Vol. 7.

Inhelder, B. *Le diagnostic du raisonnement chez les debéles mentaux*. Neuchâtel, Switzerland: Delachaux et Niestlé, 1963.

Minuchin, S. The study and treatment of families who produce multiple acting out boys. *American Journal of Orthopsychiatry* 34 (1964), 813–822.

Minuchin, S., et al. *Families of the slums: An exploration of their structure and treatment*. New York: Basic Books, 1967.

Nuffield Mathematics Project. New York: Wiley, 1968.

Piaget, J. *The construction of reality in the child*. New York: Basic Books, 1954; London: Routledge and Kegan Paul, 1955.

Piaget, J. *Language and thought of the child*. New York: International Universities Press, 1955; Neuchâtel et Paris: Delachaux et Niestlé, 1923.

Piaget, J. *La genese des structures logiques élémentaires*. Neuchâtel, Switzerland: Delachaux et Niestlé, 1959.

Piaget, J. *Play, dreams, and imitation in childhood*. New York: Norton, 1962; London: Routledge and Kegan Paul, 1951.

Piaget, J. *Judgment and reasoning in the child*. Totowa, N.J.: Littlefield, Adams, 1966(a); London: Routledge and Kegan Paul, 1928.

Piaget, J. *The psychology of intelligence*. Totowa, N.J.: Littlefield, Adams, 1966 (b); London: Routledge and Kegan Paul, 1950.

Sigel, I. The child's attainment of concepts. In M. Hoffman and L. Hoffman (eds.), *Review of child development*. New York: Russell Sage Foundation, 1964. Vol. I.

Sinclair-de-Zwart, H. Developmental psycholinguistics. In D. Elkind and J. Flavell (eds.), *Studies in cognitive development: Essays in honor of Piaget*. New York: Oxford University Press, 1968. Pp. 315–336.

CHAPTER 12

THE HAVING OF WONDERFUL IDEAS

Eleanor Duckworth

Kevin, Stephanie, and the Mathematician

Recently, I reviewed some classic Piaget interviews with a few children, to show them to a friend. One involved seriation of lengths. I had cut 10 cellophane drinking straws into different lengths and asked the children to put them in order, from smallest to biggest. The first two 7 year olds did it with no difficulty and little interest. Then came Kevin. Before I said a word about the straws, he picked them up and said to me, "I know what I'm going to do," and proceeded, on his own, to seriate them by length. He didn't mean, "I know what you're going to ask me to do." He meant, "I have a wonderful idea about what to do with these straws. You'll be surprised by my wonderful idea."

It wasn't easy for him. He needed a good deal of trial and error as he set about developing his system. But he was so pleased with himself when he accomplished his self-set task that when I decided to offer them to him to keep (10 whole drinking straws!), he glowed with joy, showed them to one or two select friends, and stored them away with other treasures in a shoe box.

The having of wonderful ideas is what I consider to be the essence of intellectual development. And I consider it the essence of pedagogy to give Kevin the occasion to have his wonderful ideas and to let him feel good about himself for having them. To develop this point of view and to indicate where Piaget fits in for me, I need to start with some autobiography, and I apologize for that, but it was a struggle of some years' duration for me to see how Piaget was relevant to schools at all.

I had never heard of Piaget when I first sat in a class of his in Paris in 1957. I had just received a bachelor's degree in philosophy, and it was as a philosopher that Piaget won me. In fact, he won me to such an extent that I spent 2 years in Geneva as a graduate student and research assistant. Then in 1962 I began to pay attention to schools, when, as a Ph. D. dropout, I accepted a job developing elementary science curriculum, and found myself in the midst of an exciting circle of educators.

The colleagues I admired most got along very well without any special knowledge of psychology. They trusted their own insights about when and how children were learning, and they were right. Their insights were excellent. Moreover, they were especially distrustful of Piaget. He had not yet appeared on the cover of *Saturday Review* or *The New York Times Magazine*, and they had their own picture of him: a severe, humorless intellectual confronting a small child with questions that were surely incomprehensible, while the child tried to tell from the look in his eyes what the answer was supposed to be. No wonder the child couldn't think straight. (More than one of these colleagues first started to pay attention to Piaget when they saw a photo of him. He may be Swiss, but he doesn't look like Calvin! Maybe he can talk to children after all.)

I myself didn't know what to think. My colleagues did not seem to be any the worse for not taking Piaget seriously. Nor, I had to admit, did I seem to be any the better. Schools were such complicated places compared with psychology labs that I couldn't find a way to be of any special help. Not only did Piaget seem to be irrelevant, I was no longer sure that he was right. For a couple of years, I scarcely ever mentioned him and simply went about the business of trying to be helpful, with no single instance, as I recall, of drawing directly on any of his specific findings.

The lowest point came when one of my colleagues gleefully showed me an essay written in a first grade by 6-year-old Stephanie. The children had been investigating capillary tubes, and were looking at the differences in the height of the water as a function of the diameter of the tube. Stephanie's essay read as follows: "I know why it looks like there's more in the skinny tube. Because it's higher. But the other is fatter, so there's the same."

My colleague triumphantly took this statement as proof that 6 year olds can reason about the compensation of two dimensions. I didn't know what to say. Of course, it should have been simple. Some 6 year olds *can* reason about compensations. The ages that Piaget mentions are only norms, not universals. Children develop at a variety of speeds. Some children develop slower and some develop faster. But I was so unsure of myself at that point, that this incident shook me badly, and all of that only sounded like a lame excuse.

I do have something else to say about that incident later. For now, I shall simply try to describe my struggle.

Even if I did believe that Piaget was right, how could he be helpful? If the main thing that we take from Piaget is that before certain ages children are unable to understand certain things—conservation, transitivity, spatial coordinates—what do we do about it? Do we try to teach the children these things? Probably not, because on the one hand Piaget leads us to believe that we probably won't be very successful at it; and on the other hand, if there is one thing we have learned from Piaget it is that children can be left to their own devices in coming to understand these notions. We don't have to try to furnish them. It took a few months before that was clear to me, but I did conclude that this was not a very good way to make use of Piaget.

An alternative might be to keep in mind the limits on children's abilities to classify, conserve, seriate, etc., when deciding what to teach them at certain ages. However, I found this an inadequate criterion. There was so much else to keep in mind. The most obvious reason, of course, was that any class of children has a great diversity of levels. Tailoring to an average level of development is sure to miss a large proportion of the children. In addition, a Piaget psychologist has no monopoly here. When trying to approximate the abilities of a group of children of a given age, able teachers like my colleagues could make as good approximations as I.

What I found most appealing was that the people with whom I was working judged the merits of any suggestion by how well it worked in classrooms. That is, instead of deciding on a priori grounds what children *ought* to know, or what they *ought* to be able to do at a certain age, they found activities, lessons, points of departure that would engage children in real classrooms, with real teachers. In their view, it was easy to devise all-embracing schemes of how science (as it was in this instance) could be organized for children, but to make things work pedagogically in classrooms was the difficult part. They started with the difficult part. A theory of intellectual development might have been the basis of a theoretical framework of a curriculum. But in making things work in a classroom, it was but a small part compared with finding ways to interest children, to take into account different children's interests and abilities, to help teachers with no special training in the subject, and so forth. So, the burden of this curriculum effort was classroom trials. The criterion was whether or not they worked, and their working depended only in part on their being at the right intellectual level for the children. They might be perfectly all right, from the point of view of intellectual demands, and yet fall short in other ways. Most often, it was a complex combination.

As I was struggling to find some framework within which my knowledge of Piaget would be useful, I found, more or less incidentally, that I was starting to be useful myself. As an observer for some of the pilot teaching of this program, and later as a pilot teacher myself, I found that I had some good insights into intellectual difficulties that children encountered. I had a certain skill in being able to watch and listen to children and figure out how they were really seeing the problem. This led to a certain ability to raise questions that made sense to the children or to think of a new orientation for the whole activity that might correspond better to their way of seeing things. I don't want to suggest that I was unique in this. Many of the excellent teachers with whom I was in contact had similar insights, as did many of the mathematicians and scientists among my colleagues, who, from their points of view, could tell when children were seeing things differently from the ways they did. But the question of whether or not I was unique is not really pertinent. For me, through my experience with Piaget of working closely with one child at a time and trying to figure out what was really in his mind, I had gained a wonderful background for being sensitive to children in classrooms. I feel that a certain amount of this kind of background would be similarly useful for every teacher.

This sensitivity to children in classrooms continued to be central in my own development. As a framework for thinking about learning, my understanding of Piaget has been invaluable. This understanding, however, has also been deepened by working with teachers and children. I may be able to shed some light on that mutual relationship by referring again to 6-year-old Stephanie's essay on compensation. Few of us, looking at water rise in capillary tubes of different diameters, would bother to wonder whether the quantities are the same. Nobody asked Stephanie to make that comparison and, in fact, it is impossible to tell just by looking. On her own, she felt it was a significant thing to comment upon. I take that as an indication that for her it was a wonderful idea. Not long before, she believed that there was more water in the tube in which the water was higher. She had recently won her own intellectual struggle on that issue, and she wanted to point out her finding to the world for the benefit of those who might be taken in by preliminary appearances.

This incident, once I had figured it out, helped me think about a point that had bothered me in one of Piaget's anecdotes. You may recall Piaget's account of a mathematician friend who inspired his studies of the conservation of numbers. This man told Piaget about an incident from his childhood, where he counted a number of pebbles he had set out in a line. Having counted them from left to right and found there were 10, he decided to see how many there would be if he counted them from right to left. Intrigued to find that there were still 10, he put them in a different arrangement and counted them again. He kept rearranging and counting them until he decided that, no matter what the arrangement, he was always going to find that there were 10. Number is independent of the order of counting.

My problem was this: in Piaget's accounts of his subjects, if 10 eggs are spread out so they take more space than 10 egg cups, a classic nonconserver will maintain that there are more eggs than egg cups, even if he counts and finds that he comes to 10 in both cases. Counting is not sufficient to convince him that there are enough egg cups for all the eggs. How is it, then, that for the mathematician, counting was sufficient? If he was a nonconserver at the time, counting should not have made any difference. If he was a conserver, he should have known from the start that it would always come out the same.

I think it must be that the whole enterprise was his own wonderful idea. He raised the question for himself and figured out for himself how to try to answer it. In essence, I am saying that he was in a transitional moment, and that Stephanie and Kevin were, too. He was at a point where a certain experience fit into certain thoughts and took him a step forward. A powerful pedagogical point can be made from this.

These three instances dramatize it because they deal with children moving ahead with Piaget notions, which are usually difficult to advance on the basis of any one experience. The point has two aspects: First, the right question at the right time can move children to peaks in their thinking that result in significant steps forward and real intellectual excitement; and, second, although it is almost impossible for an adult to know exactly the right time to ask a specific question of a specific child—especially for a teacher who is concerned with 30 or more children—children can raise the right question for themselves if the setting is right. Once the right question is raised, they are moved to tax themselves to the fullest to find an answer. The answers did not come easily in any of these three cases, but the children were prepared to work them through. Having confidence in one's ideas does not mean, "I know my ideas are right"; it means, "I am willing to try out my ideas."

As I put together experiences like these and continued to think about them, I started developing some ideas about what education could be and about the relationships between education and intellectual development.

Hank

It is a truism that all children in their first and second years make incredible intellectual advances. Piaget has documented these advances from his own point of view, but every parent and every psychologist know this to be the case. One recurring question is, why does the intellectual development of vast numbers of children then slow down? What happens to children's curiosity and resourcefulness later in their childhood? Why do so few continue to have their own wonderful ideas? I think part of the answer is that intellectual breakthroughs come to be less and less valued. Either they are dismissed as being trivial—as Kevin's or Stephanie's or the mathematician's might have been by some adults—or else they are discouraged as being unacceptable—like discovering how it feels to wear shoes on the wrong feet, or asking questions that are socially embarrassing, or destroying something to see what it's like inside. The effect is to discourage children from exploring their own ideas and to make them feel that they have no important ideas of their own, only silly or evil ones.

But I think there is at least one other part of the answer, too. Wonderful ideas do not spring out of nothing. They build on a foundation of other ideas. The following incident may help to clarify what I mean.

Hank was an energetic and not very scholarly fifth grader. His class had been learning about electric circuits with flashlight batteries, bulbs, and various wires. After the children had developed considerable familiarity with these materials, the teacher made a number of mystery boxes. Two wires protruded from each box, but inside, unseen, each box had a different way of making contact between the wires. In one box the wires were attached to a battery; in another they were attached to a bulb; in a third, to a certain length of resistance wire; in a fourth box they were not attached at all; etc. By trying to complete the circuit on the outside of a box, the children were able to figure out what made the connection inside the box. Like many other children, Hank attached a battery and a bulb to the wire outside the box. Because the bulb lit, he knew at least that the wires inside the box were connected in some way. But, because it was somewhat dimmer than usual, he also knew that the wires inside were not connected directly to each other and that they were not connected by a piece of ordinary copper wire. Along with many of the children, he knew that the degree of dimness of the bulb meant that the wires inside were connected either by another bulb of the same kind or by a certain kind of resistance wire.

The teacher expected them to go only this far. However, in order to push the children to think a little further, she asked them if they could tell whether it was a bulb or a piece of wire inside the box. She herself thought there was no way to tell. After some thought, Hank had an idea. He undid the battery and bulb that he had already attached on the outside of the box. In their place, using additional copper wire, he attached 6 batteries in a series. He had already experimented enough to know that 6 batteries would burn out a bulb, if it was a bulb inside the box. He also knew that once a bulb is burned out, it no longer completes the circuit. He then attached the original battery and bulb again. This time he found that the bulb on the outside of the box did not light. So he reasoned, with justice, that there had been a bulb inside the box and that now it was burned out. If there had been a wire inside, it would not have burned through and the bulb on the outside would still light.

Note that to carry out that idea, Hank had to take the risk of destroying a light bulb. In fact, he did destroy one. In accepting this idea, the teacher had to accept not only the fact that Hank had a good idea that even she did not have, but also that it was worthwhile to destroy a small piece of property for the sake of following through an idea. These features almost turn the incident into a parable. Without these kinds of acceptance, Hank would not have been able to pursue his idea. Think of how many times this acceptance is not forthcoming in the life of any one child.

But the main point to be made here is that in order to have his idea, Hank had to know a lot about batteries, bulbs, and wires. His previous work and familiarity with those materials were a necessary aspect of this occasion for him to have a wonderful idea. David Hawkins has said of curriculum development, "You don't want to cover a subject; you want to uncover it." That, it seems to me, is what schools should be about. They can help to uncover parts of the world which children would not otherwise know how to tackle. Wonderful ideas are built on other wonderful ideas. They do not occur contentless. In Piaget's terms, you must reach out to the world with your own intellectual tools and grasp it; assimilate it yourself. All kinds of things are hidden from us—even though they surround us—unless we know how to reach out for them. Schools and teachers can provide materials and questions in ways that suggest things to be done with them; and children, in the doing, cannot help being inventive.

There are two aspects to providing occasions for wonderful ideas. One is being willing to accept children's ideas. The other is providing a setting that suggests wonderful ideas to children—different ideas to different children—as they get caught up in intellectual problems that are real to them.

What Schools Can Do

Recently I had the chance to evaluate an elementary science program. It happened to be set in Africa, but for the purposes of this discussion it might have been set anywhere. Although the program was by no means a formal attempt to apply Piaget, it was, to my mind, an application of Piaget in the best sense. The assumptions that lay behind the work correspond well with Piaget's views of the nature of learning and intellectual development. In fact, they correspond with the ideas I have just been developing. The program set out to reveal the world to children. They sought to familiarize the children with the material world—that is, with biological phenomena, physical phenomena, and technological phenomena—flashlights, mosquito larvae, clouds, clay. When I speak of familiarity, I mean that the child should feel at home with these things: He should know what to expect of them, what can be done with them, how they react to various circumstances, what he likes about them and what he does not like about them, and how they can be changed, avoided, preserved, destroyed, or enhanced.

Certainly the material world is too diverse and too complex for a child to become familiar with all of it in the course of an elementary school career. The best that one can do is to make such knowledge, such familiarity, seem interesting and accessible to the child. That is, one can familiarize him with a few phenomena in such a way as to catch his interest, to let him raise and answer his own questions, to let him realize that his ideas are significant—so that he has the interest, the ability, and the self-confidence to go on by himself.

Such a program is a curriculum, so to speak, but a curriculum with a difference. The difference can best be characterized by saying that the unexpected is valued. Instead of expecting teachers and children to do only what was specified in the booklets, the program sought that children and teachers would have so many unanticipated ideas of their own about the materials that they would never even use the booklets. The purpose of developing booklets at all is that teachers and children start producing and following through their own ideas, if possible getting beyond needing anybody else's suggestions. This is unlikely ever to be completely realized, of course. However, as an ideal it represents the orientation of the program. It is a rather radical view of curriculum development.

It is just as necessary for teachers as for children to feel confidence in their own ideas. It is important for them as people and it is important in order for them to feel free to acknowledge the children's ideas. If teachers feel that their class must do things just as the book says and that their excellence as a teacher depends upon this, they cannot possibly accept the children's divergence and creations. A teachers' guide must give enough indications, enough suggestions, so that the teacher has ideas to start with and to pursue. But it must also enable the teacher to feel free to move in her own directions when she has other ideas.

For instance, the teachers' guides for this program include many examples of things children are likely to do. The risk is that teachers may see these as things that the children in their classes *must* do. Whether or not the children do them becomes a measure of successful or unsuccessful teaching. Sometimes the writers of the teachers' guides intentionally omit mention of some of the most exciting activities because they almost always happen even if they are not arranged. If the teacher expects them, she will often force them, and they no longer happen with the excitement of wonderful ideas. Often the writers include extreme examples, so extreme that a teacher cannot really expect them to happen in her class. These examples are meant to convey the message that "even if the children do that it's OK! Look, in one class they even did this!" This approach often is more fruitful than the use of more common examples whose message is likely to be "this is what ought to happen in your class."

The teachers' guides dealt with materials which were readily available in or out of schools, and suggested activities that could be done with these materials so that children became interested in them and started asking their own questions. For instance, common substances all around us are the basis of chemistry knowledge. They interact together in all sorts of interesting ways that are accessible to all of us if only we know how to reach out for them. Here is an instance of a part of the world waiting to be uncovered. How can it be uncovered for children in a way that gives them an interest in continuing to find out about it, that gives them the occasion to take their own initiatives and to feel at home in this part of their world?

The teachers' guide suggests starting with salt, ashes, sugar, cassava starch, alum, lemon juice, and water. When mixed together, some of these cause bubbles. Which combinations cause bubbles? How long does the bubbling last? How can it be kept going longer? What other substances cause bubbles? If a combination bubbles, what can be added that will stop the bubbling? Other things change color when they are mixed together, and similar questions can be asked of them.

Written teachers' guides, however, cannot bear the burden alone, if this kind of teaching is totally new. To get such a program started, a great deal of teacher education is necessary as well. Although I shall not try to go into this in any detail, there seem to be three major aspects to such teacher education. First, teachers themselves must learn in the way that the children in their classes will be learning. Almost any one of the units developed in this program is as effective with adults as it is with children. The teachers themselves learn through some of the units and feel what it is like to learn in this way. Second, the teachers work with one or two children at a time so that they can observe them closely enough to realize what is involved for the children. Last, it seems valuable for teachers to see films or live demonstrations of a class of children learning in this way, so that they can start to feel that it really is possible to run their class in such a way. A fourth aspect is of a slightly different nature. Except for the rare teacher who will take this leap all on her own on the basis of a single course and some written teachers' guides, most teachers need the support of at least some nearby coworkers who are trying to do the same thing, and with whom they can share notes. An even better help is the presence of an experienced teacher to whom they can go with questions and problems.

Eleanor Duckworth

An Evaluation Study

What the children are doing in one of these classrooms may be lively and interesting, but it would be helpful to know what difference the approach makes to them in the long run, to compare in some way the children who had been in this program with children who had not, and to see whether in some standard situation they now act differently.

One of my thoughts about ways in which these children might be different was based on the fact that many teachers in this program had told us that their children improved at having ideas of what to do, at raising questions, and at answering their own questions; that is, at having their own ideas and being confident about their own ideas. I wanted to see whether this indeed was the case.

My second thought was more ambitious. If these children had really become more intellectually alert, so that their minds were alive and working not only in school but outside school, they might, over a long enough period of time, make significant headway in their operational thinking, as compared to other children.

In sum, these two aspects would put to the test my notions that the development of intelligence is a matter of having wonderful ideas and feeling confident enough to try them out; and that schools can have an effect on the continuing development of wonderful ideas. Although the study has been written up elsewhere (Duckworth, 1971), I shall summarize it here.

The evaluation had two phases. The procedure developed for the first phase was inspired in part by a physics examination given to students at the Massachusetts Institute of Technology by Philip Morrison. His examination was held in the laboratory. The students were given sets of materials, the same set of materials for each student, but they were given no specific problem. Their problem was to *find* a problem and then to work on it. For Professor Morrison, the crucial thing is finding the question, just as it was for Kevin, Stephanie, and the mathematician. Indeed, in this examination, clear differences in the degree of both knowledge and inventiveness were revealed in the problems the students set themselves and the work they did was only as good as their problems.

In our evaluation study, we had to modify this procedure somewhat to make it appropriate for children as young as 6 years of age. Our general question was what children with a year or more of experience in this program would do with materials when they were left to their own resources without any teacher at all. We wanted to know whether children who had been in the program had more ideas about what to do with materials than did other children.

The materials we chose were not, of course, the same as those that children in the program had studied. We chose materials of two sorts: on the one hand, imported materials that none of the children had ever seen before—plastic color filters, geometric pattern blocks, folding mirrors, commercial buildings sets, for example. On the other hand, we chose some materials that were familiar to all the children whether or not they had been in the program—cigarette foil, match boxes, rubber rings from inner tubes, scraps of wire, wood, and metal, empty spools, and so forth.

From each class we chose a dozen children at random and told them—in their own vernacular—to go into the room and do whatever they wanted with the materials they would find there. We told them that they could move around the room, talk to each other, and work with their friends.

We studied 15 experimental classes and 13 control classes from first to seventh grade. Briefly, and inadequately, summarizing the results of this phase, we found that the children who had been in the program did indeed have more ideas about how to work with the materials. Typically, the children in these classes would take a first look at what was offered, try a few things, and then settle down to work with involvement and concentration. Children sometimes worked alone and sometimes collaborated. They carried materials from table to table, using them in ways we had not anticipated. As time went on, there was no sign that they were running out of ideas. On the contrary, their work became so interesting that we were always disappointed to have to stop them after 40 minutes.

By contrast, the other children had a much smaller range of ideas about what to do with the materials. On the one hand, they tended to copy a few leaders. On the other hand, they tended to leave one piece of work fairly soon and to switch to something else. There were few instances of elaborate work in which a child spent a lot of time and effort to overcome difficulties in what he was trying to do. In some of these classes, after 30 or 35 minutes, all the children had run out of ideas and were doing nothing.

We had assessed two things in our evaluation: diversity of ideas in a class, and depth to which the ideas were pursued. The experimental classes were overwhelmingly ahead in each of the two dimensions. This first phase of assessment was actually a substitute for what we really wanted to do. Ideally, we wanted to know whether the experience of these children in the program had the effect of making them more alert, more aware of the possibilities in ordinary things around them, and more questioning and exploring during the time they spent outside school. This would be an intriguing question to try to answer, but we did not have the time to tackle it. The procedure that we did develop, as just described, may have been too close to the school setting to give rise to any valid conclusions about what children are like in the world outside school. However, if you can accept with me, tentatively, the thought that our results might indicate a greater intellectual alertness in general—a tendency to have wonderful ideas—then the next phase takes on a considerable interest.

I am hypothesizing that this alertness is the motor of development in operational thinking. No doubt there is a continuum. No normal child is completely unalert. But some are far more alert than others. I am also hypothesizing that a child's alertness is not fixed. By opening up to children the many fascinating aspects of the ordinary world and by enabling them to feel that their ideas are worthwhile having and following through, I believe that their tendency to have wonderful ideas can be affected in significant ways. This program seemed to be doing both those things, and by the time I evaluated it, some children had been in the classes for up to 3 years. It seemed to me that we might—just might—find that the 2 or 3 years of increased alertness that this program fostered had made some difference to the intellectual development of the children.

In the second phase, then, we examined the same children individually, using Piaget problems administered by a trained assistant who spoke the language of the children. A statistical analysis revealed that on 5 of the 6 problems we studied, the children in the experimental classes did significantly better than the children in the comparison classes.

I find this to be a stunning result. It is the only program I know even to suggest that something happening in schools might make a difference to operational development.

However, by no means do I suggest that a goal for education is the acceleration of Piaget stages. It is the theoretical aspect that I find important. My thesis at the outset of this chapter was that the development of intelligence is a matter of having wonderful ideas. In other words, it is a creative affair. When children are afforded the occasions to be intellectually creative—by being offered matter to think about and by having their ideas accepted—then not only do they learn about the world, but as a happy side effect their general intellectual ability is stimulated as well.

Another way of putting this is that I think the distinction made between "divergent" and "convergent" thinking is oversimple. Even to think a problem through to its most appropriate end-point (convergent) one must create various hypotheses to check out (divergent). When Hank came up with a closed end-point to the problem, it was the result of a brilliantly imaginative—that is, divergent—thought. We must conceive of the possibilities before we can check them out.

Conclusion

I am suggesting that children do not have a built-in pace of intellectual development. I would temper that suggestion by saying that the built-in aspect of the pace is minimal. The having of wonderful ideas, which I consider the essence of intellectual development, would depend instead to an overwhelming extent on the occasions for having them. I have dwelt at some length on how important it is to allow children to accept their own ideas and to work them through. I would like now to consider the intellectual basis for new ideas.

I react strongly against the thought that we need to provide children with only a set of intellectual processes—a dry, contentless set of tools that they can go about applying. I believe that the tools cannot help developing once children have something real to think about; and if they don't have anything real to think about, they won't be applying tools anyway. That is, there really is no such thing as a contentless intellectual tool. If a person has some knowledge at his disposal, he can try to make sense of new experiences and new information related to it. He fits it into what he has. By knowledge I do not mean verbal summaries of somebody else's knowledge. I do not urge a return to textbooks and lectures. I mean a person's own repertoire of thoughts, actions, connections, predictions, and feelings. Some of these may have as their source something he has read or heard. But he has done the work of putting them together for himself, and they give rise to new ways for him to put them together.

The greater the child's repertory of actions and thoughts—in Piaget's terms, schemes—the more material he has for trying to put things together in his head. The essence of the African program I described is that children increase the repertories of actions that they carry out on ordinary things, which in turn gives rise to the need to make more intellectual connections.

Let us consider a child who has had the world of common substances opened to him, as described earlier. He now has a vastly increased repertory of actions to carry out and of connections to make. He has seen that when you boil away sea water, a salt residue remains. Would some residue remain if he boiled away beer? If he dissolved this residue in water again, would he have beer again—flat beer? He has seen that he can get a colored liquid from flower petals if he crushes them. Could he get that liquid to go into water and make colored water? Could he make colored coconut oil this way? All these questions and the actions they lead to are based on the familiarity the child has gained with the possibilities contained in this world of common substances.

Intelligence cannot develop without matter to think about. Making new connections depends on knowing enough about something in the first place to provide a basis for thinking of other things to do—of other questions to ask—that demand more complex connections in order to make sense. The more ideas a person already has at his disposal about something, the more new ideas occur and the more he can coordinate to build up still more complicated schemes.

Piaget has speculated that some people reach the level of

formal operations in some specific area that they know well—auto mechanics, for example—without reaching formal levels in other areas. That fits into what I am trying to say. In an area you know well, you can think of many possibilities, and working them through demands formal operations. If there is no area in which you are familiar with the complexities enough to work through them, then you are not likely to develop formal operations. Knowing enough about things is one prerequisite for wonderful ideas.

I shall make one closing remark. The wonderful ideas that I refer to need not necessarily look wonderful to the outside world. I see no difference in kind between wonderful ideas that many other people have already had, and wonderful ideas that nobody has yet happened upon. That is, the nature of creative intellectual acts remains the same, whether it is an infant who for the first time makes the connection between seeing things and reaching for them, or Kevin who had the idea of putting straws in order of their length, or a cook who conceives of a new combination of herbs, or an astronomer who develops a new theory of the creation of the universe. In each case, new connections are being made between things already mastered. The more we help children to have their wonderful ideas and to feel good about themselves for having them, the more likely it is that they will some day happen upon wonderful ideas that no one else has happened upon before.

REFERENCE

Duckworth, E. A comparison study for evaluating the Primary School Science in Africa: African Primary Science Program. Newton, Mass.: 1971 (unpublished manuscript).

CHAPTER 13

THE DEVELOPING TEACHER

Milton Schwebel and Jane Raph

Listening to discussions about highly structured curricula one is struck by the implication that teachers need an instruction manual that tells them in detail how to select the lesson, what to do next, what questions to ask, what answers to obtain, and what replies to give. We may be born free, but we are not free agents in the classroom, at least not if the proponents of the machine-tending model have their way.

One opportunity a teacher must have is free use of her intelligence. Lacking that, she is no boon to any classroom of children no matter what other qualities she has. Surely that is not a surprising assertion in a book devoted to the flowering of intelligence in children. Yet, it needs to be said again and again to counter the strong emphasis on the kind of curriculum development that is "idiot-free," that is, so constructed that even an idiot could use it and get good results.

There is nothing in the work of Piaget nor in the chapters of this book that suggests there ought to be Piaget schools as there were "Progressive Schools." His work (and that of others) does, however, make a strong stand for schools whose curricula, particularly at the elementary school level, would simply be occasions for developing the mind. Obviously, during these occasions the child will learn as a by-product all kinds of facts and skills, some of them highly important, but the emphasis will be on the intelligence, not on the facts and skills. Thus there are Piaget schools only in the sense that teachers benefit from the knowledge of half a century of research, and that principals value and support teachers whose behavior is guided by that knowledge.

The implications for the principal, the consultant, and the teacher-educator are clear: to behave in such ways as to encourage teachers to do their own thinking and rely on their own judgement in the unpredictable life of the classroom. Chittenden,[1] referring to his observations as advisor-consultant in Follow Through, said that coming into the school as the expert defeats the purpose of trying to help the teachers get a broader outlook toward their children.

> A better approach is to try very truly to respect the teacher for where she may be at that point in her own teaching and to really mean this. . . . The approach insofar as possible is to treat the teacher in the way one would like the teacher to treat the child, and that is with respect and with interest in the original contributions of the teacher.
>
> Toward the end after our visits we drew the general conclusion that there should be two kinds of questions that one asks when visiting a school. One is, to what extent are the children moving into a kind of a central decision-making role? The other question, not independent of the first, is to what extent is there movement of the adult from a role of a curriculum follower to a thinking, mature behaving, resourceful adult? The approach of the advisor-consultant should be to identify a starting point with the teacher and to work with that and not come in inadvertently or consciously proclaiming a whole set of generalizations (Piagetian or otherwise) which a teacher "should" use.

As Chittenden points out, it is not an easy matter for consultants, principals, educators, or teachers themselves to change because they have all been through very much the same schooling that generally rewards the bland acceptance of what is taught—the scholastic orthodoxy. And those who incorporate it most thoroughly reap the benefits in school and college. Because we are not accustomed to another way and find it alien or discomforting when we first encounter it, even when our prior learning did not help us very much we go back for more of the same. Furth[2] has reported:

Many teachers both at the University and in schools do not know how to ask for help, to put it briefly. Often what they seem to be asking for are just the kinds of things that our experience tells us cannot or ought not be given. I get requests to "teach a course," to "provide the newest information on linguistic knowledge . . . knowledge that teachers should know, linguistic generalizations that can be applied . . . etc." I know that if I were to comply with such requests, it would not be very useful to them. A year afterward they will know that all the generalizations that I provided for them did not help very much.

Passivity is the enemy of intellectual and social development. It is everybody's enemy and sits like a Trojan horse, variously taking the form of the well-behaved child, the good follower, the good listener, and the permissive parent or teacher. The effort against it must be energetic and unrelentless if the mind is to be free to act upon the environment of the school. It is less important to concern ourselves with the accumulation of a body of information than it is to sensitize teachers to the way children think and to the way thinking develops. Furth[3] tells what he did with teachers in one school system in which he served as consultant:

While working with one group of teachers, I demonstrated the concept of probability by playing a probability game with about fifteen children. We used a box into which I put ten yellow blocks, two blue blocks, and one green block. I put the blocks in the box in front of the children so they did not have to memorize what the box contained. . . . ten yellow, two blue, and one green block. Then I questioned each child asking "Take out a block after guessing what color you think it will be." The children appeared to be excited by the task and were enjoying it. Some were guessing, some were guessing blindly, some were using some strategy. They reasoned the block extracted would be yellow because there were ten, or blue because there were two. When the block extracted was actually green, they thought it was magic.

After this demonstration we talked about the probability concept. My point was not to give them the probability concept, not to teach them a strategy, but to give them an opportunity to work it out. We worked with the children for a twenty minute period each week for a month. Those children who were ready did very well. This may be the occasion on which they articulated their probability concept. However, again, the point was not to teach the children (or the teachers) the probability concept, but to give them experiences which would allow them to construct the idea.

We also talked about dramatic play as a way of encouraging thinking. We had a child play a certain age. For instance, we asked an eight year old child to sit on a chair and wait for a bus, and to act a certain age. The other children had to evaluate his performance and make an intelligent guess as to what age the child was playing. Now I'm not a dramatic coach. I'm sure I didn't handle this too well, but the children became very involved. The teachers were impressed at the possibilities of this approach. They said, "Of course we can do this." There are many ideas that children can act out which they cannot articulate in words or which they cannot write about.

The developing teacher, as Furth indicates, needs to understand the child's mind in order to choose appropriate teaching procedures and to make the many impromptu decisions that are part of life in a classroom. As several authors have noted, what the teacher teaches is not always assimilated by the child in the intended fashion, because the child can absorb only what he can accommodate. Frequently he deforms the information and it comes out on examinations in conclusions that are hardly recognizable. Sometimes the child shows a mechanical and mindless attitude toward that which he has been formally taught, yet a considerable use of thought and judgment about related matters in which he has not had formal classroom work. Ginsburg,[4] discussing research in collaboration with Robert Davis of the Madison Project and Dr. Ezra Heitowit of Cornell University, comments at length about children displaying that contradictory behavior.

What does the child learn . . . in school? Let us limit this to a particular academic subject, mathematics. A parent, for instance, wants to know what children, in fact, acquire and she is not satisfied with the usual concept of it, namely with the notion that they acquire some quantum of arithmetic ability or that their number concepts somehow increase in mass. She wants to know much more about what they know in some detail. To get some such preliminary idea, I have been conducting some unstructured Piagetian clinical interviews, with children concerning their knowledge of mathematics taught in school.

One thing that children engage in when they do school mathematics is computation. And this is usually considered a very trivial aspect of their work on mathematics. It is usually assumed that the child learns the algorithm that the textbook teaches. One thing that we find is that this isn't necessarily the case. We asked one child how much seven and seven is. He did not know; then he started to think about it and then he said, "Well, six plus six is twelve, plus two is fourteen." Now, this was not the way he was taught to do it. Let me give another example. A child was given a problem, something like 13 into 84. And he said, "Well, maybe 13 goes into 84, 3 times." Actually he said, "Maybe 84 goes into 13, 3 times." He was a little confused. He thought a little while and I said, "How do you know?" He wrote down his answer, and he said, "Well, that's wrong." He added 13 and he said, "That's wrong." "Plus 13?" and so on, and so on, until he found the right answer. The point about this is that when a child is doing calculations in school he does not necessarily do what the teacher is trying to teach him to do. He is in a way assimilating the problem, the computation, into various frameworks that he has. That is to say, he knows how much six and six is, he's memorized that and so he makes use of that fact in approaching this new problem of computation. This child knew how to add pretty well, he evidently had difficulty with multiplication. So he transformed multiplication into repeated addition which actually indicates a good deal of insight into the nature of multiplication. If we consider computation, the computational system of the child—we are not talking now about his understanding of mathematics or anything; we are talking about computation skills—these turn out to be far more elaborate, from a cognitive point of view, then we would have supposed.

Let us pursue this further by seeing another computational case. We asked a girl of 8 or 9 to add some numbers. We said, "Add 23 and 482." She wrote it down this way:

$$\begin{array}{c} 23 \\ \underline{482} \end{array}$$

"Well, 2 plus 4 is 6, 3 plus 8 is 11. You carry the 1. And that's that." We were interested in this. She had a set of rules for doing this, very sensible set of rules. It happens to run off backwards. We gave her other series of numbers and she wrote them the same way, lining them up wrong. We asked her to add them up and she went through the same algorithm as before, starting from the left, carrying over here, and so on. New this is again an example of a computational system that is lawful, and regular but wrong, in this case. The next step is, we say, "Well, suppose you had to guess how much this is."

```
14,887
 2,036
   427
   242
   131
   101
    64
    27
     9
______
```

She says, "Well, about 18,000." This took her a little bit of time. And then we said, "Well, how did you do that?" And she said, "Well, there's about 15,000 here, and about 2,000 there, that's 17,000 and all this is about 1,000. So, it would be about 18,000." And then she says, "Well, maybe 18,341." Now, this is not calculation, in the ordinary sense. This is a heuristic; this is a way of obtaining an intelligent guess. Now again, this heuristic has a structure. The structure is something like this. Take the largest number, drop off all the little irrelevant things toward the end, take the next largest number, drop off all the little irrelevant things. I've got 15,000 plus 2,000, that's easy to add. Everything else you can lump together. We did not really pursue this heuristic that much. We could describe it in more detail than I have just done. But you can see it is a regular, sensible procedure.

We have seen other heuristics that children have used that are analogous to this. The child develops heuristics many times independently of what the teacher would like the child to learn, because teachers do not like to proximate answers, they want to know exactly what it is. The child has to buck the system, to a certain extent, to develop sensible heuristics. What is the structure of these heuristics? It is an interesting question because if you have a sensible heuristic, it is based on a lot of knowledge of mathematics. This child knows, for example, that the sum has to be greater than the part, in positive numbers, at least. She knows that largest numbers in a sequence like that make up the greater part of the sum. She knows that it is safe to ignore these small numbers, more or less. Now that is considerable insight into mathematics, for a girl who adds like this. On a computational level she is incredibly stupid, on the heuristic level, she is very bright. And that heuristic is based on some very fundamental ideas concerning addition.

There is another interesting thing about the problem the girl had with addition; at first she did not recognize the conflict. How come you got different results, we asked her. Why did you line things up this way? And then eventually, she had the thought that went something like this: "Oh, I've got the place values done badly." When that idea of place value which was part of her formal knowledge of mathematics was applied to this isolated computational system, she had this "Ah, ha" insight. Then she revised the whole computational system. And that again is another common thing . . . we find in kids working mathematics, there is a kind of schizophrenia between different intellectual systems which are isolated from one another. And when you begin to break some of the barriers, then some of these things straighten out.

Ginsburg's teaching theory and method, developed more fully in a recent book (1972), are consistent with Sinclair's procedure reported in Chapter 3 on facilitating transition to a higher operational level: Ask the kinds of questions that enable the child to see the contradictions. If she has learned about place values but only in a formal or mechanical fashion, confront her with the alternatives. Using her algorithm, 482 plus 1 is 582; the alternative answer is 483. Anyone interested in children's *thinking* as opposed to their memorizing a rule would want her to experience the contradiction.

It seemed as if this girl had been taught that when she added she ignored her thinking operations and did as she was told; whereas, if she were asked to have a guess—if she could not fall back on some memorized response—then she could think again. Perhaps one of the major things for teachers to learn from this is a very simple idea: Rules for doing things can be applied in a stupid fashion that covers up abilities that are there.

Written communication poses special problems for children and their teachers. Unlike the mathematical area that, for all of the individualized thought of the child and the variations with stages, still has objectively correct answers, that of writing has so much that is subjective that it is more difficult to study. Parker[5] wrestles with some of the problems of teachers of English.

My background is almost entirely in English—English teaching, and teaching English teachers. This past year I was running a year-long program for eight experienced English teachers who had spent anywhere from three to about twenty years teaching English in Chicago public high schools.

These people suffered, not only as students in schools, but also as teachers, from having been taught that their own personal experiences were not valid data for making inferences about themselves, the world, about teaching, about whatever they are involved in. And one thing I was interested in doing was finding out if there were some kinds of experiences that I could provide in a university setting that would, in a sense, put these people back in touch with their own best intuitions, their own best sense about what they should be doing with kids in schools.

My own greatest interest is in writing and the teaching of writing, so that is the point of entry that I chose. The sequence evolved, not really as one that I planned beforehand, but as one that was decided quickly. First of all, all eight people were involved in working in some kind of new art form: painting, working with clay, something like that. They were making something in a new form, not one that they had worked in before, or at least had not worked in for a good number of years. As part of this process, they were asked to record their own process at various stages, to keep some kind of written account of what they went through in making things in new forms.

The session turned out to be a kind of child-like experience for them. There were even comments like, "Gee, I feel like a five year old or a six year old . . . trying to paint."

The second step was then to move to a writing workshop. I put them through a sequence of writing exercises: not only were they to write specific things, but again they were to keep a running journal on their own writing process. The third step was to have them read comments practicing writers had written about their own writing processes.

With these three sources of data, two of them pretty much here and now and experiential, one less so (the comments of various people: Henry Miller, Norman Mailer, people who have, in fact, made some comments on their own writing process), I then asked them to consider the typical things that they had been doing for some years, in teaching writing to children. Very quickly, nearly every one of them in some way came to see that their practices, methods, strategies for the teaching of writing were irrelevant, or antagonistic, to the ways in which pieces of writing, their own and others, were made.

The instrumental metaphor for teaching writing during the whole history of English teaching has been a think-write metaphor. One thinks about something, then one inscribes that thought on the paper. Now, that metaphor works quite nicely if the kind of writing you are talking about is logical, expository, writing. But it does not work for other modes of writing.

The think-write metaphor, and the kinds of pedagogy that teachers set around it, are adult perceptions based on adult abstractions. Probably there is a unique rhetoric of children's writing, just as there are unique qualities in children's thinking.

Let me seem to contradict myself for a moment and say that, in most instances anyway, writing is a reflection of some thought process. But I am not sure that it is always just a translation of that process, or a transcription of some fully formulated thought. In other words, I really believe that there is some kind of discovery, or invention, that happens in the writing act. For example, young children produce writing in forms that they do not know about, and this writing often points to ideas that they have not "thought about" in any formal sense. I had seen kids write drama even though they have never read a play . . . and make very sophisticated dramaturgical decisions in their writing, even though they don't know anything about dramaturgy. In fact, they probably would not have made the decisions had the teacher begun by saying, "Now we are going to learn something about dramaturgy, and then we are going to write plays."

I really do believe that this whole business of writing is developmental. I have another isolated bit of evidence. A group of ninth grade Basic Track students at a high school in Chicago, about twenty youngsters, fifteen and sixteen years of age, all of whom were reading below fifth grade level, were asked simply to record for about 20 minutes what they were experiencing of internal and external events. Examining what they produced, I was struck by the incredible ease with which they could handle this task. They were perceptive about what they were experiencing, outside—what was going on in the room, what they were hearing and seeing and smelling. They were also very fluent in recording what they were feeling, what they were thinking, so much so that without telling this group of experienced teachers that I had seen the kids' writing, I asked them to do the same kind of thing in our class. And, to a person, they were almost unable to do it. They really could not handle it. Maybe in some developmental sense, here was a particular part of writing operation, maybe a kind of mental operation, that certain people aged fifteen could handle that adults, these adults at least, ranging in age from 27 to 43 could not handle. And these were all teachers who had been teaching these kinds of kids, who were going back to teach these kinds of kids. It was very striking for them to find out that they could not. It was unsettling and gave them a sharper notion of the appropriateness of assignments.

It was good for teachers with whom Parker worked to be shaken by the experience of discovering that their "disadvantaged" students succeeded where they could not. It was good because experiences that shake the teachers' equilibrium offer some promise of opening new vistas to them. For both teachers and children, situations that conflict with their conceptions evoke responses necessary to regain a balance. Piaget posits a self-regulating process that enables us to adapt to changes, be they external or internal ones. In the process of adaptation, that is, by the active steps of adaptation, intelligence develops. Behavior that is more intelligent can follow the confrontation of contradictions. These can be the teachers' recognition that the backward adolescents possess a competency that they lack or that their own capacity to observe the world about them has been dulled; or it can be the child's recognition that her method of addition yields a patently incorrect sum.

The joyous state that some writers put forth as educational utopia seems more like a refuge for happy morons. Contradictions and disequilibria are necessary to intellectual growth. They instigate adjustments that lead to the accommodation of higher levels of knowledge. The tension that accompanies such contradictions is inescapable; in fact, it ought to be welcomed when the child or adult gets those first intimations that all is not well with the way he is understanding some phenomenon. "Yes, it does seem strange that the quantity of water that appears to me to be greater in the tall, thin glass than in the small, wide one has the same water. Something is wrong there." Or, "I never could see why John got so angry when I took the toy from him because I wasn't angry with him. But could it be that he feels the way I feel when someone takes the toy from me?" Or, "I added those numbers just the way the teacher taught us. I memorized it and used it that way. Now, why does the answer seem so wrong?" These questions are meant to convey the experience but not the conscious thought of children at the point at which contradictions emerge.

It is highly similar with adults. Developing teachers, sensing that all is not well with the functioning of their class, will begin to question their orientation and procedures. It is a well-known fact, documented recently in dozens of books, that many new teachers respond to their tension-provoking recognition by changing only one variable—the degree of strictness. The teacher becomes "tougher" but otherwise remains unchanged and closes off the opportunity, for the time being at least, of developing by means of a new adaptation under circumstances of disequilibrium. That is, the teacher's understanding of how a good teacher behaves in class has been undermined by experience. Either the teacher maintains the same posture in the belief that it needs to be accompanied by greater firmness, or the teacher changes it. Whichever choice is made, a new equilibrium is established, providing the decision leads to a reduction of tension and a greater compatibility between the teacher's understanding of how she should be functioning and the outcome of that functioning.

No one can deny that any group requires "discipline" if it is going to be productive in achieving the end it sets for itself. A poker game, a sensitivity training group, a family, or a class of adults or children each requires some degree of regulation, self- or otherwise determined. When that discipline breaks down, the immediate survival-type response of the leader frequently is to seek to impose his authority all the more. Under certain conditions that action may turn out to be the appropriate response, although these are probably few in number and occur in groups characterized by considerable internal dissension and the potential for disruption. In most instances, however, that reaction does not work if "working" means the improvement of education, for even if it should succeed in reducing the tensions of the teacher it might be a highly inappropriate response when things are not going well in class. The reason might be that a group of students, particularly those who do not have high extraneous motivation, has not been engaged. Boring and repetitive activity—even worse, meaningless content in the sense that the children do not comprehend it—might be responsible for the state of affairs that has brought the teacher to the necessity for changed behavior. The best hope for the teacher is to do for her students what she likes done for herself in university classes: to be intellectually engaged, challenged, and excited, as much as the infant is when he explores a wooden block or a toy, or as a scientist is when he studies the atom. Teachers will argue that they must cope with a "meat and potato" business and that they must have enough peace and quiet in the classroom to get this underway. True, but the sociopsychological climate in the classroom is not unrelated to the teacher's philosophy of instruction: One does not take precedence over the other, because they are interdependent.

The developing teacher discovers the power of the action-oriented approach and finds it effective with all age levels including adults. Naturally, the future teacher who does herself experience that kind of instruction with professors who provide challenge and confrontations, not simply predigested knowledge, is likely to incorporate an inquiring approach into her way of life, consequently into her teaching. This approach is important not only because of the quality and style of teaching but also because of the adult model the teacher represents. Our point is that the application of Piagetian theory to the classroom has widespread ramifications. It offers the promise of facilitating the development of intelligence, at the same time as it engages the interest of the students and concomitantly promotes good conditions and climate for learning and teaching—that is, for experiences that sharpen the contradictions in mental structures rather than in interpersonal conflicts. The disequilibria can thereby be productive. Lest there be any misinterpretation of the statement about the possibilities inherent in the application of the theories, we must say that these are aspirations toward which many of us are striving, through experience and research. The studies and observations of Chittenden, Furth, Ginsburg, and Parker are testimony to the search for answers which is entirely in accord with the Piagetian theoretical underpinnings for pedagogy. It is a search and a striving, as it is for the developing teacher who works out an effective style suited to his or her own personality and way of relating to people.

A well-known marine biologist and a Nobel laureate, Szent-Gyorgyi (1964) has expressed very clearly a problem with which most of us as teachers have to contend:

> There is a widely spread misconception about the nature of books which contain knowledge. It is thought that such books are some-

thing the contents of which have to be drummed into our heads. I think the opposite is closer to the truth. Books are there to keep the knowledge in while we use our heads for something better (p. 1278).

Obviously this biologist does not propose that we use our heads for thinking *in vacuo*, which is not the alternative to brainstuffing about our academic fields. He means (as the authors of this book mean) using our heads in connection with any and all of the fields of knowledge that we study. Using our heads about mathematics, literature, chemistry, and music is, in this view, the optimal way of studying these subjects. In one sense, this means a change in the subject matter one studies because, as one aims to develop intelligence in the process of teaching subject *X*, the substance of *X* changes by virtue of the changed objectives and the resultant change in procedures or methodology.

In this book the focus has been on teachers, children, and the development of the mind, with only occasional reference to but no analysis of a wider social system. Sarason (1971) has rightly pointed to the predominance of psychological studies of the schools, and in particular, studies that concentrate on the individual and individual difference, and to the neglect of an ecological approach. Our lack of emphasis does not at all mean any low valuation. On the contrary, a more holistic approach to the application of the Piagetian, or any other theory that marks a break with orthodoxy in school practice, requires careful consideration be given to the culture of the school and the political and social forces that determine its character. The fate of innovation rests there.

NOTES

1. Chittenden, Edward. *Symposium: Problems, issues, questions, and application of Piaget's Theory*. Conference, Application of Piagetian Theory to Education: An Inquiry Beyond the Theory. Rutgers University, Graduate School of Education, New Brunswick, New Jersey, 1970.

2. Furth, H. *Symposium*, Ibid.
3. Ibid.
4. Ginsburg, H. *Symposium*, op. cit.
5. Parker, R. *Symposium*, op. cit.

REFERENCES

Ginsburg, H. *The myth of the deprived child: Poor children's intellect and education*. Englewood Cliffs, N.J.: Prentice-Hall, Inc., 1972.

Sarason, S. *Culture of the school and the problem of change*. Boston: Allyn and Bacon, 1971.

Szent-Gyorgyi, A. Teaching and the expanding knowledge. *Science 146*, 4 (December, 1964).

INDEX